T0340188

STATISTICAL
THERMODYNAMICS OF
SEMICONDUCTOR ALLOYS

STATISTICAL THERMODYNAMICS OF SEMICONDUCTOR ALLOYS

VYACHESLAV A. ELYUKHIN

Centro de Investigacion y de Estudios Avanzados del Instituto Politecnico Nacional, Mexico

ELSEVIER

AMSTERDAM • BOSTON • HEIDELBERG • LONDON • NEW YORK • OXFORD
PARIS • SAN DIEGO • SAN FRANCISCO • SINGAPORE • SYDNEY • TOKYO

Elsevier
Radarweg 29, PO Box 211, 1000 AE Amsterdam, Netherlands
The Boulevard, Langford Lane, Kidlington, Oxford OX5 1GB, UK
225 Wyman Street, Waltham, MA 02451, USA

Notices
Knowledge and best practice in this field are constantly changing. As new research and
experience broaden our understanding, changes in research methods, professional
practices, or medical treatment may become necessary.

Practitioners and researchers must always rely on their own experience and knowledge
in evaluating and using any information, methods, compounds, or experiments
described herein. In using such information or methods they should be mindful of their
own safety and the safety of others, including parties for whom they have a professional
responsibility.

To the fullest extent of the law, neither the publisher nor the authors, contributors, or
editors, assume any liability for any injury and/or damage to persons or property as a
matter of products liability, negligence or otherwise, or from any use or operation of any
methods, products, instructions, or ideas contained in the material herein.

ISBN: 978-0-12-803987-8

British Library Cataloguing in Publication Data
A catalogue record for this book is available from the British Library

Library of Congress Cataloging-in-Publication Data
A catalog record for this book is available from the Library of Congress

For information on all Elsevier publications
visit our website at http://store.elsevier.com/

Dedication

To Lyuda

Contents

Preface

Due to their electronic structure and optical properties, semiconductors are the basic materials of solid-state electronics. Among semiconductor materials, crystalline (polycrystalline) inorganic semiconductor alloys form the majority of the materials used in device applications. The doped elemental semiconductors and doped semiconductor compounds such as, correspondingly, doped Si or doped GaAs, are also substitutional semiconductor alloys. Moreover, a number of the semiconductor compound-rich alloys containing dielectric compounds are semiconducting materials.

Over the last few years, progress in the technologies of epitaxial growth led to significant extension of the class of inorganic semiconductor alloys. It is expected that this extension will be continued in the next years. In connection with this trend, the prediction of the characteristics and properties of possible semiconductor materials becomes very important. In addition, the materials suitable to fabricate solid-state electronic devices should be in the thermodynamically stable or metastable state. Only then can the characteristics of the devices be fixed for a long period of time. Therefore, the development of methods to determine the thermodynamic stability of semiconductor alloys with respect to phase transformations is very important.

The electronic structure and optical properties are considered in great detail in a number of books devoted to the physics of semiconductors. The thermodynamic properties and characteristics (e.g., the thermodynamic stability with respect to the different phase transformations, clustering, and distortions of the crystal structure) are also essential for device fabrication. However, these and other thermodynamic quantities, characteristics, and properties are normally represented only briefly in the available literature. Some models and their applications suitable for the consideration of the thermodynamic properties of semiconductor alloys can be found, for example, in Refs [1–5] and in the books devoted to statistical mechanics and statistical thermodynamics.

This book is for solid-state physicists, semiconductor materials scientists, and specialists who need the application of theoretical methods and models, such as the use of the lattice systems, the cluster variation method, regular solutions, and the valence force field models to analyze the experimental results and to predict the properties and characteristics of possible semiconductor alloys. Both graduate and postgraduate students of solid-state physics and materials science may use this book to study the methods of statistical physics in solid-state physics as well as the regular solution model and the valence force field model and their applications to the description of semiconductor alloys.

The topics presented in this book include the types of inorganic crystalline semiconductors, the basic concepts and postulates of equilibrium thermodynamics and equilibrium statistical physics, the regular solution model and its applications considered by using the cluster variation method, and the valence force field model with its applications. Elemental semiconductors, semiconductor compounds, and substitutional alloys of such semiconductors are treated in Chapter 1. The most important parts of this chapter are devoted to the alloys of binary semiconductor compounds. The established types of the spontaneously ordered alloys with the zinc blende and wurtzite structures are described. It is shown that in alloys with two mixed sublattices, a one-to-one correspondence is absent between the elemental composition and concentration of chemical bonds. The basic concepts and the mathematical formalism of equilibrium thermodynamics, as well as main elements of statistical physics, are briefly treated in Chapter 2. In addition, the Helmholtz and Gibbs free energies of condensed matter and the separation of the degrees of freedom are presented. The classical regular solution model is in Chapter 3. High accuracy in calculation of configurational entropy may be reached by using this model. The cluster variation method and its different approximations used to describe the properties and characteristics of the semiconductor alloys considered as regular solutions are in Chapter 4. Baker's approach to providing a simple and systematic way to express configurational entropy is introduced and used in this chapter. Chapter 5 is devoted to the modified regular solution model to describe the semiconductor alloys of binary compounds in which the crystal structure consists of two mixed sublattices. A one-to-one correspondence between the elemental composition and concentration of chemical bonds is absent in such alloys. The characteristics of such alloys also represented as regular solutions are treated by using the different approximations of the cluster variation method. It is also shown that the self-assembled identical tetrahedral clusters should be thermodynamically profitable in some semiconductor alloys with the zinc blende structure. The valence force field model and its applications are presented in Chapter 6. This model allows analyzing the distortions of the crystal structure of the mismatched semiconductor alloys at the microscale level. The estimated internal strain energies of the semiconductor alloys with the zinc blende and wurtzite structures demonstrate the tendency to the formation of superstructures. The possible types of superstructures in the ternary alloys with the zinc blende and wurtzite structures are described. The models of the discontinuous and continuous order–disorder phase transitions are presented. The strain energies caused by the isoelectronic impurities in the semiconductors with the diamond and zinc blende structures are derived.

In this book, the terms in parentheses normally are the synonyms encountered in the literature, and capital and small letters as usual are related to the absolute and molar quantities, respectively.

References

[1] R.A. Swalin, Thermodynamics of Solids, John Wiley & Sons, New York, 1962, 1972.

[2] J.C. Phillips, Bonds and Bands in Semiconductors, Academic Press, New York, 1973.

[3] J. Tsao, Materials Fundamentals of Molecular Beam Epitaxy, Academic Press, Boston, 1993.

[4] A.-B. Chen, A. Sher, Semiconductor Alloys, Plenum Press, New York, 1995.

[5] L.A. Girifalco, Statistical Mechanics of Solids, Oxford University Press, Oxford, 2000.

Semiconductor Materials

Crystalline inorganic semiconductors are basic materials of solid-state electronics and, therefore, they are considered in this book. An inorganic semiconductor can be an elemental semiconductor, a compound, or an alloy. Elemental semiconductors containing only one chemical element are diamond, Si, Ge, and gray Sn with the diamond structure. Semiconductor compounds consist of two or more chemical elements. The composition of the compound may vary within a composition range called the deviation from stoichiometry or deviation from the ideal ratio between the numbers of the different chemical elements. Normally, such deviations in semiconductor compounds are small, and they are the results of the presence of atoms located at the interstitials or the absence of atoms over the lattice sites. Semiconductor alloys are mainly substitution alloys in which atoms are arranged over the lattice sites. Interstitials and the absence of atoms over the lattice sites also may occur in semiconductor alloys as defects of the crystal structure. The importance of the consideration of substitutional semiconductor alloys results from the fact that they are basic materials in the device applications.

The compressibility of crystalline semiconductor alloys is normally very small. Therefore, the changes of the lattice parameters of semiconductors are significant only at high pressures. The high-pressure effects are not considered in this book. Hence, through the book it is supposed that the lattice parameters of semiconductors do not depend on pressure. Moreover, it is supposed also that the stiffness coefficients of semiconductors do not depend on temperature and the coefficients of thermal expansion are equal to zero for the majority of considerations.

1.1 ELEMENTAL SEMICONDUCTORS

Carbon (diamond, C), silicon (Si), germanium (Ge), and gray tin (α-Sn) are the elemental semiconductors belonging to Group IV of the periodic

Statistical Thermodynamics of Semiconductor Alloys
http://dx.doi.org/10.1016/B978-0-12-803987-8.00001-0 **1**

FIGURE 1.1 Elemental semicon-
ductors A^{IV} with the diamond structure.

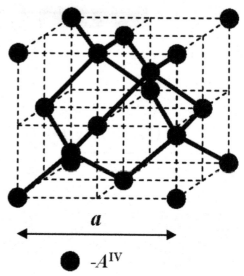

a

● - A^{IV}

table. They have the diamond structure (cubic structure). The elementary
cell (unit cell) of the diamond structure is shown in Figure 1.1.

The diamond structure is composed of two equivalent face-centered
cubic lattices displaced from each other by one-quarter of a body diagonal. Also, the diamond structure can be represented as a set of regular
tetrahedrons (tetrahedral cells) with atoms in their corners, and 50% of
such tetrahedrons have atoms in their centers. Other tetrahedrons are
empty.

The Bravais lattice of the diamond structure is the face-centered cubic
lattice. Each atom in such a structure has the four nearest neighbors with
the tetrahedral bonding and the twelve next nearest neighbors (the
nearest neighbors in the face-centered cubic lattice). The distance between
the nearest atoms is:

$$R = \frac{\sqrt{3}}{4}a,$$

where a is the lattice parameter or length of the edge of the elementary
cell.

1.2 SEMICONDUCTOR COMPOUNDS WITH ZINC BLENDE STRUCTURE

Binary compounds consisting of atoms belonging to Groups III and V
of the periodic table ($A^{III}B^{V}$ semiconductors) such as BP, BAs, AlP, AlAs,
AlSb, GaP, GaAs, GaSb, InP, InAs, and InSb crystallize with the zinc

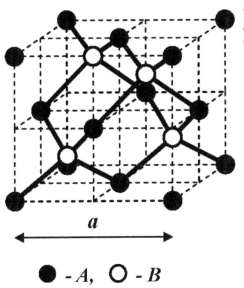

FIGURE 1.2 *AB* semiconductor compounds with the zinc blende structure.

a

● - *A*, ○ - *B*

blende (sphalerite) structure. The elemental cell of the zinc blende structure is shown in Figure 1.2.

The other $A^{III}B^V$ semiconductors such as BN, AlN, GaN, and InN can be grown with the zinc blende structure in the thermodynamically metastable state. $A^{II}B^{VI}$ semiconductors BeS, BeSe, BeTe, MgSe, ZnS, ZnSe, ZnTe, CdS, CdTe, HgSe, and HgTe have the zinc blende structure in the thermodynamically stable state. The other $A^{II}B^{VI}$ semiconductors MgS, ZnO, and CdSe can be prepared with the zinc blende structure in the thermodynamically metastable state. The $A^{IV}B^{IV}$ compound SiC also may be grown with the zinc blende structure in the metastable state.

The zinc blende structure is obtained from the diamond structure if cations are placed into one face-centered cubic sublattice and anions are allocated into another face-centered cubic sublattice. There are four atoms of the opposite type placed at the corners of a regular tetrahedron around each atom. Thus, the nearest coordination number z_1 is equal to 4 and the next nearest coordination number z_2 is equal to 12. The next nearest neighbors are the same type atoms as a central atom. The Bravais lattice of the zinc blende structure is the face-centered cubic lattice as well as the Bravais lattice of the diamond structure. All distances between the nearest atoms in the zinc blende structure are the same, and given by:

$$R = \frac{\sqrt{3}}{4}a.$$

where a is the lattice parameter or length of the edge of the elementary cell. As well as the diamond structure, the zinc blende structure can be represented as a set of regular tetrahedrons (tetrahedral cells) with cations (anions) in their corners and 50% of such tetrahedrons have anions (cations) in their centers. The other tetrahedrons are empty.

1.3 SEMICONDUCTOR COMPOUNDS WITH WURTZITE STRUCTURE

$A^{III}B^V$ semiconductors AlN, GaN, and InN crystallize with the wurtzite structure. $A^{II}B^{VI}$ semiconductors BeO, MgTe, ZnO, CdS, and CdSe also crystallize with the wurtzite structure in the most stable state. $A^{II}B^{VI}$ compounds ZnS and ZnSe can be grown with the wurtzite structure in the thermodynamically metastable state. The $A^{IV}B^{IV}$ semiconductor SiC crystallizes normally with the wurtzite structure.

The ideal wurtzite structure (Figure 1.3) consists of two hexagonal close-packed sublattices filled with cations and anions and displaced from each other by a distance $\sqrt{\frac{3}{8}}a_0$, where a_0 is the first lattice parameter of the ideal hexagonal close-packed sublattice or the distance between the nearest neighbors in the hexagonal close-packed sublattice in the (0001)

FIGURE 1.3 *AB* semiconductor compounds with the wurtzite structure.

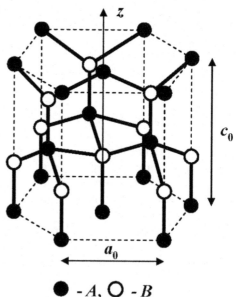

$\bullet$ - A, $\bigcirc$ - B

plane. The ratio between the first and second lattice constants of the ideal hexagonal close-packed sublattice is given by $\frac{c_0}{a_0} = \sqrt{\frac{8}{3}} = 1.633.$

The relative displacement $c_0 u_0$ of two hexagonal close-packed sublattices in the ideal wurtzite structure is represented by a parameter $u_0 = 3/8 = 0.375$, which is the ratio between the distance between the nearest neighbors along the z axis and the lattice parameter c_0. As well as the zinc blende structure, the wurtzite structure can be represented as a set of regular tetrahedrons (tetrahedral cells) with cations (anions) in their corners, and 50% of such tetrahedrons have anions (cations) in their centers. The other 50% of the tetrahedrons are empty.

Normally, the structures of the compounds crystallized with the wurtzite structure are slightly distorted from the ideal wurtzite structure, so that the ratio $\frac{c}{a} < \frac{c_0}{a_0}$ and the parameter $u > 0.375$ if such compounds are in the thermodynamically stable state. Accordingly, the structures of the compounds in the metastable state that can be grown with the wurtzite structure are also slightly distorted, so that the ratio $\frac{c}{a} > \frac{c_0}{a_0}$. Accordingly, the distance between the nearest neighbors along the axis z is different from the length of the other bonds.

Each atom is at the center of a regular tetrahedron formed by the atoms of another sublattice and it has 12 next nearest neighbors of the same type of atoms. The Bravais lattice of the wurtzite structure is the simple hexagonal lattice. Thus, the wurtzite structure consists of four Bravais lattices.

1.4 SEMICONDUCTOR COMPOUNDS WITH ROCK SALT STRUCTURE

$A^{II}B^{VI}$ compounds MgO, MgS, CaO, CaS, SrO, SrS, BaO, BaS, and CdO, and $A^{IV}B^{VI}$ compounds such as PbS, PbSe, and PbTe crystallize with the rock salt (sodium chloride) structure. The elemental cell of the rock salt structure shown in Figure 1.4 consists of the cationic and anionic face-centered cubic sublattices displaced from each other by one-half of the edge of a cube in the sublattice.

The Bravais lattice is the face-centered cubic lattice. The nearest and next nearest neighbors are equal to 6 and 12, respectively. The distance between the nearest neighbors is:

$$R = \frac{a}{2},$$

where a is the lattice parameter.

FIGURE 1.4 *AB* semiconductor compounds with the rock salt structure.

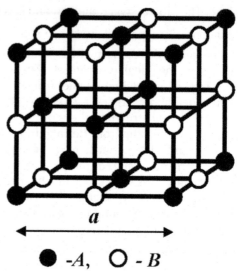

a

● -A, ○ - B

1.5 SEMICONDUCTOR COMPOUNDS WITH CHALCOPYRITE STRUCTURE

Many $A^IB^{II}C^{IV}_2$ (A^I = Cu, Ag; B^{II} = Al, Ga, In, Tl; C^{IV} = S, Se, Te) ternary compounds crystallize with the chalcopyrite structure. There are also many $A^{II}B^{IV}C^V_2$ (A^{II} = Mg, Zn, Cd; B^{IV} = Si, Ge, Sn; C^V = N, P, As) ternary compounds that crystallize with the chalcopyrite structure or undergo the phase transformation from the state with the zinc blende structure in the disordered state to the state with the chalcopyrite structure in the ordered state. The chalcopyrite structure consists of two geometrically identical cation sublattices and one anion sublattice. One of the cation sublattices fills with A atoms and atoms B form another cation sublattice. There are two geometrically equivalent sublattices filled with cations A and B and anions C, respectively, in the disordered state of ABC_2 compounds. The elemental cell of the chalcopyrite structure is shown in Figure 1.5.

The ideal chalcopyrite structure (typical for CuInTe$_2$, ZnSnP$_2$, and ZnSnAs$_2$ ternary compounds) consists of two regular cubes in the ordered state. However, the majority of ternary compounds with the chalcopyrite structure exhibit the tetragonal distortion in which $c/a < 2$ in the ordered state. The transition from the chalcopyrite structure to the zinc blende structure can be made as follows. One of the face-centered cubic sublattices is anionic formed by atoms belonging to Group IV or Group V of the periodic table in $A^IB^{II}C^{IV}_2$ and $A^{II}B^{IV}C^V_2$ compounds, respectively. The cationic face-centered cubic sublattice consists of two geometrically equivalent sublattices filled with Group I and

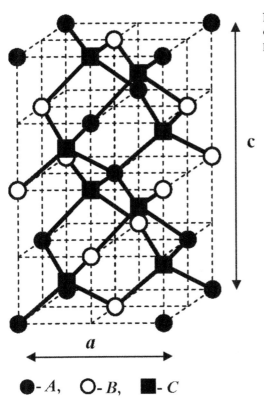

FIGURE 1.5 ABC_2 semiconductor compounds with the chalcopyrite structure.

$\bullet$ - A, $\bigcirc$ - B, $\blacksquare$ - C

Group II atoms or with Group II and Group IV atoms in $A^{I}B^{II}C^{IV}_2$ and $A^{II}B^{IV}C^{V}_2$ compounds, respectively. The Bravais lattice of the chalcopyrite structure is body centred tetragonal.

1.6 ALLOYS OF ELEMENTAL SEMICONDUCTORS

Among alloys of the elemental semiconductors, Ge_xSi_{1-x} substitutional alloys with the diamond structure are the most studied. Ge_xSi_{1-x} are alloys since chemical bonds between the different nearest neighbors do not form therein. The elemental cell of Ge_xSi_{1-x} alloys is shown in Figure 1.6. Normally, Ge and Si atoms randomly occupy the lattice sites and, therefore, Ge_xSi_{1-x} are substitutional alloys.

However, Ge_xSi_{1-x} alloys can be also grown atomically ordered or, in other words, with the ordered distribution of atoms over the lattice sites, or with the superstructure or with the long-range order [1]. A superstructure or a long-range order in crystalline alloys is the long-distance ordered arrangement of atoms depending on their types. The distribution of atoms

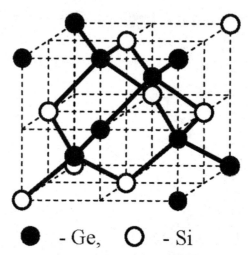

$\bullet$ - Ge, $\bigcirc$ - Si

FIGURE 1.6 Ge_xSi_{1-x} alloys with the diamond structure.

in atomically ordered Ge_xSi_{1-x} alloys has a double lattice periodicity in the $\langle 111 \rangle$ direction. Thus, Ge and Si atoms are preferentially in comparison with the disordered state allocated in the pairs of planes perpendicular to the $\langle 111 \rangle$ direction. The structure of completely ordered $Ge_{0.5}Si_{0.5}$ alloy with the pairs of planes filled with Ge and Si atoms is shown in Figure 1.7.

This type of atomic ordering is called the double CuPt-type super-structure. The name occurred due to a similarity to the CuPt super-structure in which there are alternating (111) planes preferentially occupied by Cu and Pt atoms in the face-centered cubic crystal lattice of Cu_xPt_{1-x} alloy. The double CuPt-type superstructure in Ge_xSi_{1-x} alloy is formed as a result of surface kinetics of physisorbed atoms during the non-equilibrium epitaxial growth on the Si (001) substrates. Ge_xSi_{1-x} alloys can be formed as thermodynamically stable disordered alloys at their melting temperatures for any composition but they are a miscibility gap at the lower temperatures.

Carbon-rich Si_xC_{1-x}, Ge_xC_{1-x}, and Sn_xC_{1-x} semiconductors are not considered here since the solubility of Si, Ge, and Sn in diamond is negligibly small. As for Si-rich alloys, Si-rich C_xSi_{1-x} alloys are most likely the Si matrix with embedded 1C4Si complexes that are tetrahedral cells with carbon central atoms due to the large cohesive energy of compound SiC. We can assume so, since carbon atoms form the chemical bonds with Si atoms in Si-rich C_xSi_{1-x} alloys. In Ge-rich C_xGe_{1-x} alloys, the majority of carbon atoms are over the lattice sites. Therefore, such alloys can be considered as substitutional alloys with the diamond structure. The Si-rich Sn_xSi_{1-x} alloys and Ge-rich Sn_xGe_{1-x} alloys crystallize with the diamond structure and disordered allocation of atoms over the lattice

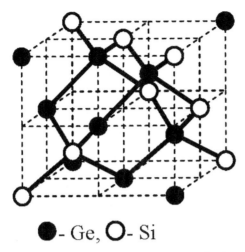

$\bullet$ - Ge, $\bigcirc$ - Si

FIGURE 1.7 $Ge_{0.5}Si_{0.5}$ alloy with the double completely ordered CuPt-type superstructure.

sites. The ternary Ge- and Si-rich $Sn_xGe_ySi_{1-x-y}$ alloys have also the diamond structure and disordered atomic arrangement.

1.7 TERNARY ALLOYS OF BINARY COMPOUNDS

Two types of atoms in ternary alloys of two binary compounds may be Group II, Group III, or Group IV atoms, and the third type of atoms should be Group V or Group VI atoms in $A^{II}{}_xB^{II}{}_{1-x}C^{VI}$, $A^{III}{}_xB^{III}{}_{1-x}C^{V}$, and $A^{IV}{}_xB^{IV}{}_{1-x}C^{VI}$ alloys, and, vice versa, in $A^{II}B^{VI}{}_xC^{VI}{}_{1-x}$, $A^{III}B^{V}{}_xC^{V}{}_{1-x}$, and $A^{IV}B^{VI}{}_xC^{VI}{}_{1-x}$ alloys. Thus, two types of atoms fill one sublattice of the crystal lattice and atoms of the third type form another sublattice. The sublattice consisting of two types of atoms is called a mixed sublattice. The nearest neighbors of atoms of the mixed sublattice are atoms of another sublattice or atoms of one type in the alloys with the zinc blende, wurtzite, and rock salt structures. The types of the pairs of the nearest neighbors correspond to two binary compounds constituting the alloy. Therefore such alloys are called the alloys of two binary compounds. There are two types of the concentrations that can be used for ternary semiconductor alloys of binary compounds. The concentrations of the first type are the concentrations of chemical bonds between the nearest neighbors or concentrations of compounds or the chemical composition of an alloy. Therefore, a ternary alloy of two binary compounds is written as $A_xB_{1-x}C$ or AB_xC_{1-x}. The unit cells of $A_xB_{1-x}C$ or AB_xC_{1-x} atomically disordered alloys with the zinc blende structure are shown in Figure 1.8(a) and (b), where x is the concentration of atoms A in the

(a)   (b)

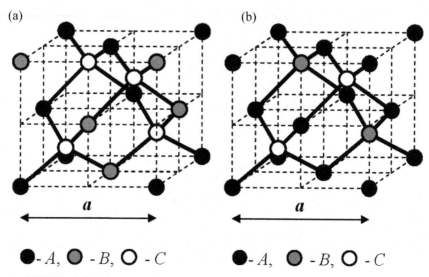

$\bullet$- A, $\oslash$ - B, $\bigcirc$ - C $\bullet$- A, $\oslash$ - B, $\bigcirc$ - C

FIGURE 1.8   (a) $A_xB_{1-x}C$ and (b) AB_xC_{1-x} alloys with the zinc blende structure.

cation-mixed sublattice of $A_xB_{1-x}C$ alloys and the concentration of B atoms in the anion-mixed sublattice of AB_xC_{1-x} alloys, respectively.

Thus, the concentration x in $A_xB_{1-x}C$ or AB_xC_{1-x} alloys is the concentration of AC or AB chemical bonds. Normally, this type of concentration is used to consider ternary alloys of binary compounds. The second type of concentration is the atomic concentrations or the elemental composition of the alloys that is given by $A_{x/2}B_{(1-x)/2}C_{1/2}$ or $A_{1/2}B_{x/2}C_{(1-x)/2}$.

The crystal structure of a ternary alloy of binary compounds is the structure of constituent compounds if the compounds have the same crystal structure. In another case, the crystal structure of an alloy corresponds to the crystal structure of the compound having the larger concentration.

$A_xB_{1-x}C$ and AB_xC_{1-x} alloys are also called quasi-binary alloys from the chemistry standpoint, since there are two types of chemical bonds in them. The crystal structure of such alloys consists of the sets of four chemical bonds of one type for alloys with the zinc blende and wurtzite structures and sets of six chemical bonds of one type for alloys with the rock salt structure. These quadruples and sextuples of bonds are around atoms forming the mixed sublattice. Such sets of chemical bonds formally can be considered as molecules of the constituent binary compounds. Accordingly, in this case, there is a one-to-one correspondence between the concentrations of atoms (elemental composition) and concentrations of chemical bonds (chemical composition). This one-to-one correspondence for $A_xB_{1-x}C$ alloys is written as $x_{AC}=x$, $x_{BC}=1-x$ and for AB_xC_{1-x} alloys is given by $x_{AB}=x$, $x_{AC}=1-x$. Thus, the concentrations of chemical bonds in ternary alloys of binary compounds are independent on the arrangement of atoms in the mixed sublattice. In other words, a change in the arrangement of atoms in the mixed sublattice does not affect the concentrations of chemical bonds.

For the first time, the atomically ordered structure in ternary alloys of two binary compounds was established in $Al_xGa_{1-x}As$ epitaxial films grown on the (100) and (110) oriented GaAs substrates [2]. The exposed atomic ordering is the preferential allocation of Al and Ga alternating monolayer (100) planes. This type of ordering of Al and Ga atoms is termed the CuAu-I superstructure due to a similarity to the CuAu-I superstructure in Cu_xAu_{1-x} alloys having the face-centered cubic crystal structure. The structure of the $Al_{0.5}Ga_{0.5}As$ completely atomically ordered alloy with the alternating (100) planes filled with Al and Ga atoms is shown in Figure 1.9. This superstructure may be considered as a structure consisting of 2Al2Ga1As tetrahedral cells in which As atoms are in their centers.

The second type of atomic ordering in alloys with the zinc blende structure was established in $GaAs_xSb_{1-x}$ and $Ga_xIn_{1-x}As$ [3,4]. This type of ordering is similar to the structure of chalcopyrite. Completely ordered $A_{0.5}B_{0.5}C$ alloy with the chalcopyrite superstructure is shown in Figure 1.10. This superstructure also can be considered as consisting of 2A2B1C tetrahedral cells in which atoms C are in their centers.

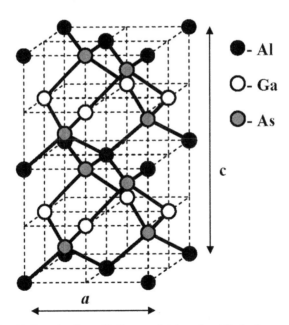

FIGURE 1.9 $Al_xGa_{1-x}As$ alloy with the completely ordered CuAu-I superstructure in the cation sublattice.

FIGURE 1.10 $A_{0.5}B_{0.5}C$ alloy with the completely ordered chalcopyrite superstructure in the cation sublattice.

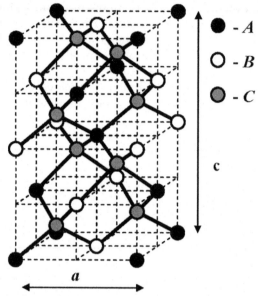

● - A

○ - B

◑ - C

The third type of atomic ordering in semiconductor alloys with the zinc blende structure is found in many alloys. The first results on this CuPt-type ordering in the mixed sublattices were published in 1987 [5–7]. The completely CuPt ordered $In_{0.5}Ga_{0.5}P$ alloys are shown in Figure 1.11(a) and (b).

The crystal structure of completely ordered $In_{0.5}Ga_{0.5}P$ is a set of 3In1Ga1P (50%) and 1In3Ga1P (50%) tetrahedral cells. The CuPt-type ordering in the III-V ternary alloys is presented by two sets of alternating $(1\bar{1}1)$ or $(\bar{1}11)$ atomic planes of the mixed sublattice filled preferentially by the atoms of the different types.

The first type of ordering in ternary alloys of the Group III nitride compounds with the wurtzite structure is represented by two sets of alternating (0001) atomic planes of the mixed sublattice consisting preferentially of the cations of the different types [8,9]. The crystal structure of such $A_{0.5}Ga_{0.5}N$ ($A = $ Al, In) completely ordered alloys shown in Figure 1.12 consists of 3A1Ga1N (50%) and 1A3Ga1N (50%) tetrahedral cells.

This type of ordering is called the WC superstructure, since the compound WC crystallizes with the hexagonal close-packed structure in which alternating (0001) atomic planes are occupied by tungsten and carbon atoms.

The other type of the superstructure in wurtzite $In_xGa_{1-x}N$ epitaxial films grown on the nonpolar $(11\bar{2}0)$ plane GaN was also identified [10]. This ordering in the wurtzite $In_xGa_{1-x}N$ films is represented by two sets

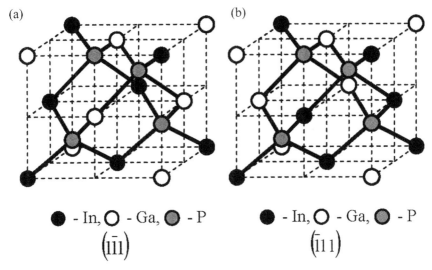

(a)

(b)

● - In, ○ - Ga, ◉ - P

● - In, ○ - Ga, ◉ - P

$(1\bar{1}\bar{1})$

$(\bar{1}1\,1)$

FIGURE 1.11 (a and b) $In_{0.5}Ga_{0.5}P$ alloys with the completely ordered CuPt superstructures in the cation sublattices.

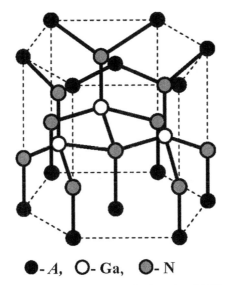

●- *A*, ○- **Ga**, ◉- **N**

FIGURE 1.12 $A_xGa_{1-x}N$ alloy with the completely ordered WC superstructures in the cation sublattices.

of alternating $(11\bar{2}0)$ atomic planes of the mixed sublattice consisting mainly of different types of atoms. The crystal structure of such completely ordered wurtzite $In_{0.5}Ga_{0.5}N$ shown in Figure 1.13 should involve only 2In2Ga1N tetrahedral cells.

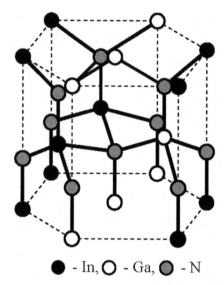

● - In, ○ - Ga, ◉ - N

FIGURE 1.13 $In_{0.5}Ga_{0.5}N$ alloy with complete ordering in the $(11\bar{2}0)$ planes.

1.8 QUATERNARY ALLOYS OF THREE BINARY COMPOUNDS

In quaternary alloys of three binary compounds, three types of atoms are placed in the mixed sublattice and atoms of the fourth type form another sublattice. These alloys are written as $A_xB_yC_{1-x-y}D$ or $AB_xC_yD_{1-x-y}$, where x is the concentration of atoms A or atoms B in the mixed sublattice, correspondingly, y is the concentration of atoms B or atoms C in the same sublattice, respectively. There are three types of chemical bonds in such quaternary alloys. Therefore, they are called quaternary alloys of three binary compounds or quasiternary alloys, since chemical bonds correspond to three binary compounds (chemical substances) in them. Each atom of the mixed sublattice of $A_xB_yC_{1-x-y}D$ or $AB_xC_yD_{1-x-y}$ alloy is surrounded by the nearest neighbors of one type in the same manner as in $A_xB_{1-x}C$ or AB_xC_{1-x} ternary alloys of binary compounds. $A_xB_yC_{1-x-y}D$ alloy with the zinc blende structure is shown in Figure 1.14.

In the case of alloys with the zinc blende and wurtzite structures, the crystal structure of quaternary alloys of three binary compounds consists of the sets of four one-type chemical bonds, and consists of the sets of six one-type chemical bonds in the case of alloys with the rock salt structure. Such sets of chemical bonds can be formally considered as molecules of the constituent binary compounds. Thus, there is a one-to-one correspondence between the concentrations of atoms (elemental composition)

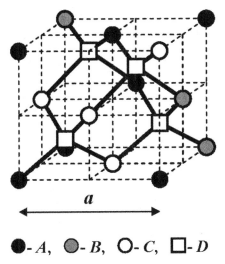

$$\bullet\text{-}A, \quad \text{\textcircled{}}\text{-}B, \quad \text{O}\text{-}C, \quad \square\text{-}D$$

FIGURE 1.14 $A_xB_yC_{1-x-y}D$ alloy with the zinc blende structure.

and concentrations of chemical bonds (chemical composition) in such alloys. Therefore, the crystal structure of quaternary alloys of three binary compounds can be represented as a structure consisting of molecules of three constituent binary compounds. The one-to-one correspondence between the concentrations of atoms and concentrations of chemical bonds is written as $x_{AD} = x$, $x_{BD} = y$, $x_{CD} = 1 - x - y$ for $A_xB_yC_{1-x-y}D$ alloys and for $AB_xC_yD_{1-x-y}$ alloys as $x_{AB} = x$, $x_{AC} = y$, $x_{AD} = 1 - x - y$. Thus, these quaternary alloys are similar to ternary alloys of binary compounds. Accordingly, the concentrations of chemical bonds in quaternary alloys of three binary compounds are independent of the arrangement of atoms in the mixed sublattice. In other words, a change of the arrangement of atoms in the mixed sublattice does not affect the concentrations of the chemical bonds in these quaternary alloys.

1.9 QUATERNARY ALLOYS OF FOUR BINARY COMPOUNDS

The quaternary alloys of four binary compounds are alloys in which each sublattice consists of atoms of two types. The cation sublattice consists of atoms belonging to the II, III, or IV Groups of the periodic table in alloys of $A^{II}B^{VI}$, $A^{III}B^V$, and $A^{IV}B^{VI}$ binary compounds, respectively. Atoms belonging to Groups VI and V form the anion sublattice. This kind of quaternary alloys is written $A_xB_{1-x}C_yD_{1-y}$. $A_xB_{1-x}C_yD_{1-y}$ alloy with the zinc blende structure is shown in Figure 1.15.

FIGURE 1.15 $A_xB_{1-x}C_yD_{1-y}$ alloy with the zinc blende structure.

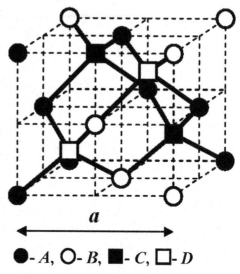

$$\bullet - A, \quad \bigcirc - B, \quad \blacksquare - C, \quad \square - D$$

Such alloys contain four types of chemical bonds AC, AD, BC, and BD and, therefore, can be considered as alloys of four chemical substances (AC, AD, BC, and BD binary compounds) corresponding to the chemical bonds. Accordingly, $A_xB_{1-x}C_yD_{1-y}$ alloys are called the quaternary alloys of AC, AD, BC, and BD binary compounds or quasiquaternary alloys. The presence of four types of the chemical bonds in these quaternary alloys provides the absence of a one-to-one correspondence between the concentrations of the atoms (elemental composition) and concentrations of the chemical bonds (chemical composition). Indeed, there are only three equations for the concentrations of the atoms and concentrations of the chemical bonds, given by:

$$x_{AC} + x_{AD} = x, \quad x_{AC} + x_{BC} = y, \quad x_{AC} + x_{AD} + x_{BC} + x_{BD} = 1.$$

These three equations have a vast number of solutions for the concentrations of the chemical bonds for one elemental composition satisfying condition x, $y \neq 1$, 0. Therefore, a one-to-one correspondence between the elemental composition of such alloy and its chemical composition is absent since one elemental composition corresponds to a vast number of chemical compositions. Any atom of $A_xB_{1-x}C_yD_{1-y}$ alloy can have atoms of two types in its nearest surroundings. Accordingly, the crystal structure of quaternary alloy of this type cannot be considered as a set of molecules of constituent compounds. The numbers of the chemical bonds are determined by the arrangement of atoms over the lattice sites. It is due to the fact that an exchange of the lattice sites between atoms belonging to the same sublattice (cations A, B or anions C, D) results in the variation of the numbers of the chemical bonds if atoms participating in the exchange have the different nearest surroundings. The variation of the numbers of the chemical bonds after the exchange corresponds to the reaction between the bonds, given by:

$$nAC + nBD \rightarrow nAD + nBC, \quad (n = 1, ..., z_1)$$

or vice versa, where z_1 is the nearest coordination number. Thus, there is the indetermination in the chemical composition of $A_xB_{1-x}C_yD_{1-y}$ alloy at the given elemental composition. This fact adds complexity to the consideration of such alloys. The indetermination in the chemical composition for alloys with the zinc blende and wurtzite structures can be illustrated as follows. The exchanges of the lattice sites between atoms from the same sublattice having the different nearest surroundings and the reactions between the chemical bonds are written as:

$$n = 1:$$

Exchange of cations

$$1A4C + 1B3C1D \rightarrow 1B4C + 1A3C1D,$$

$$1A3C1D + 1B2C2D \rightarrow 1B3C1D + 1A2C2D,$$

$$1A2C2D + 1B1C3D \rightarrow 1B2C2D + 1A1C3D,$$

$$1A1C3D + 1B4D \rightarrow 1B1C3D + 1A4D,$$

Exchange of anions

$$1C4A + 1D3A1B \rightarrow 1D4A + 1C3A1B,$$

$$1C3A1B + 1D2A2B \rightarrow 1D3A1B + 1C2A2B,$$

$$1C2A2B + 1D1A3B \rightarrow 1D2A2B + 1C1A3B,$$

$$1C1A3B + 1D4B \rightarrow 1D1A3B + 1C4B,$$

where the reaction between bonds is $AC + BD \rightarrow AD + BC$;

$$n = 2:$$

Exchange of cations

$$1A4C + 1B2C2D \rightarrow 1B4C + 1A2C2D,$$

$$1A3C1D + 1B1C3D \rightarrow 1B3C1D + 1A1C3D,$$

$$1A2C2D + 1B4D \rightarrow 1B2C2D + 1A4D,$$

Exchange of anions

$$1C4A + 1D2A2B \rightarrow 1D4A + 1C2A2B,$$

$$1C3A1B + 1D1A3B \rightarrow 1D3A1B + 1C1A3B,$$

$$1C2A2B + 1D4B \rightarrow 1D2A2B + 1C4B,$$

where the reaction between bonds is $2AC + 2BD \rightarrow 2AD + 2BC$;

$$n = 3:$$

Exchange of cations

$$1A4C + 1B1C3D \rightarrow 1B4C + 1A1C3D,$$

$$1A3C1D + 1B4D \rightarrow 1B3C1D + 1A4D,$$

Exchange of anions

$$1C4A + 1D1A3B \rightarrow 1D4A + 1C1A3B,$$

$$1C3A1B + 1D4B \rightarrow 1D3A1B + 1C4B,$$

where the reaction between bonds is $3AC + 3BD \rightarrow 3AD + 3BC$;

$$n = 4:$$

Exchange of cations

$$1A4C + 1B4D \rightarrow 1B4C + 1A4D$$

Exchange of anions

$$1C4A + 1D4B \rightarrow 1D4A + 1C4B,$$

where the reaction between bonds is $4AC + 4BD \rightarrow 4AD + 4BC$, $1A4C$ denotes atom A surrounded by four atoms C or tetrahedral cell with the central atom A, and $1A4C + 1D3A1B \rightarrow 1C3A1B + 1D4A$ is the exchange of atoms A and B between $1A4C$ and $1D3A1B$.

For $A_xB_{1-x}C_yD_{1-y}$ alloys with the rock salt structure, the exchanges of the lattice sites between cations A and B situated in the centers of the octahedral cells and reactions between chemical bonds are given by:

$$n = 1:$$

Exchange of cations

$$4C1A + 3C1D1B \rightarrow 4C1B + 3C1D1A,$$

$$3C1D1A + 2C2D1B \rightarrow 3C1D1B + 2C2D1A,$$

$$2C2D1A + 1C3D1B \rightarrow 2C2D1B + 1C3D1A,$$

$$1C3D1A + 4D1B \rightarrow 1C3D1B + 4D1A,$$

Exchange of anions

$$4A1C + 3A1B1D \rightarrow 4A1D + 3A1B1C,$$

$$3A1B1C + 2A2B1D \rightarrow 3A1B1D + 2A2B1C,$$

$$2A2B1C + 1A3B1D \rightarrow 2A2B1D + 1A3B1C,$$

$$1A3B1C + 4B1D \rightarrow 1A3B1D + 4B1C,$$

where the reaction between bonds is $AC + BD \rightarrow AD + BC$;

$$n = 2:$$

Exchange of cations

$$4C1A + 2C2D1B \rightarrow 4C1B + 2C2D1A,$$

$$3C1D1A + 1C3D1B \rightarrow 3C1D1B + 1C3D1A,$$

$$2C2D1A + 4D1B \rightarrow 2C2D1B + 4D1A,$$

Exchange of anions

$$4A1C + 2A2B1D \rightarrow 4A1D + 2A2B1C,$$

$$3A1B1C + 1A3B1D \rightarrow 3A1B1D + 1A3B1C,$$

$$2A2B1C + 4B1D \rightarrow 2A2B1D + 4B1C,$$

where the reaction between bonds is $2AC + 2BD \rightarrow 2AD + 2BC$;

$$n = 3:$$

Exchange of cations

$$4C1A + 1C3D1B \rightarrow 4C1B + 1C3D1A,$$

$$3C1D1A + 4D1B \rightarrow 3C1D1B + 4D1A$$

Exchange of anions

$$4A1C + 1A3B1D \rightarrow 4A1D + 1A3B1C,$$

$$3A1B1C + 4B1D \rightarrow 3A1B1D + 4B1C,$$

where the reaction between bonds is $3AC + 3BD \rightarrow 3AD + 3BC$;

$$n = 4:$$

Exchange of cations

$$4C1A + 4D1B \rightarrow 4C1B + 4D1A$$

Exchange of anions

$$4A1C + 4B1D \rightarrow 4A1D + 4B1C,$$

where the reaction between bonds is $4AC + 4BD \rightarrow 4AD + 4BC$, $4C1A$ denotes atom A surrounded by four atoms C or tetrahedral cell with the central atom A, and $4C1A + 3A1B1D \rightarrow 3A1B1C + 4A1D$ is the exchange of atoms A and B between $4C1A$ and $3A1B1D$ tetrahedral cells.

For $A_x B_{1-x} C_y D_{1-y}$ alloys with the rock salt structure, the exchanges of the lattice sites between cations A and B situated in the centers of the octahedral cells and reactions between chemical bonds are given by:

$$n = 1:$$

$$6C1A + 5C1D1B \rightarrow 6C1B + 5C1D1A,$$

$$5C1D1A + 4C2D1B \rightarrow 5C1D1B + 4C2D1A,$$

$$4C2D1A + 3C3D1B \rightarrow 4C2D1B + 3C3D1A,$$

$$3C3D1A + 2C4D1B \rightarrow 3C3D1B + 2C4D1A,$$

$$2C4D1A + 3C3D1B \rightarrow 2C4D1B + 3C3D1A,$$

$$1C5D1A + 6D1B \rightarrow 1C5D1B + 6D1A,$$

$$\text{reaction: } AC + BD \rightarrow AD + BC;$$

$$n = 2:$$

$$6C1A + 4C2D1B \rightarrow 6C1B + 4C2D1A,$$

$$5C1D1A + 3C3D1B \rightarrow 5C1D1B + 3C3D1A,$$

$$4C2D1A + 2C4D1B \rightarrow 4C2D1B + 2C4D1A,$$

$$3C3D1A + 1C5D1B \rightarrow 3C3D1B + 1C5D1A,$$

$$2C4D1A + 6D1B \rightarrow 4C2D1B + 6D1A,$$

$$\text{reaction: } 2AC + 2BD \rightarrow 2AD + 2BC;$$

$$n = 3:$$

$$6C1A + 3C3D1B \rightarrow 6C1B + 3C3D1A,$$

$$5C1D1A + 2C4D1B \rightarrow 5C1D1B + 2C4D1A,$$

$$4C2D1A + 1C5D1B \rightarrow 4C2D1B + 1C5D1A,$$

$$3C3D1A + 6D1B \rightarrow 3C3D1B + 6C1A,$$

$$\text{reaction: } 3AC + 3BD \rightarrow 3AD + 3BC;$$

$$n = 4:$$

$$6C1A + 2C2D1B \rightarrow 6C1B + 2C2D1A,$$

$$5C1D1A + 1C5D1B \rightarrow 5C1D1B + 1C5D1A,$$

$$4C2D1A + 6D1B \rightarrow 4C2D1B + 6D1A,$$

$$\text{reaction: } 4AC + 4BD \rightarrow 4AD + 4BC;$$

$$n = 5:$$

$$6C1A + 1C5D1B \rightarrow 6C1B + 1C5D1A,$$

$$5C1D1A + 6D1B \rightarrow 5C1D1B + 6D1A,$$

$$\text{reaction: } 5AC + 5BD \rightarrow 5AD + 5BC;$$

$$n = 6:$$

$$6C1A + 6D1B \rightarrow 6C1B + 6D1A + 6C1B,$$

$$\text{reaction: } 6AC + 6BD \rightarrow 6AD + 6BC,$$

where $6C1A$ is the octahedral cell with the central atom A. The reactions describing the exchanges of the lattice sites between anions C and D can be considered in a similar way. Cations A and B should be substituted for anions C and D, respectively, and vice versa.

1.10 QUATERNARY ALLOYS OF TERNARY COMPOUNDS

There are three types of quaternary alloys with the chalcopyrite structure denoted by $A_xD_{1-x}BC_2$, $AD_xB_{1-x}C_2$ and $ABD_{2x}C_{2(1-x)}$. $A_xD_{1-x}BC_2$ and $AD_xB_{1-x}C_2$ alloys are crystallographically similar since the cation sublattices in the chalcopyrite structure are identical geometrically. There are three types of cation–anion chemical bonds in $A_xD_{1-x}BC_2$ and $AD_xB_{1-x}C_2$ alloys in which three types of cations and the chemical bond concentrations are determined by the elemental composition due to one type of anions. The bond concentrations for $A_xD_{1-x}BC_2$ alloys are given by:

$$x_{AC} = 0.5x, \quad x_{DC} = 0.5(1-x) \text{ and } x_{BC} = 0.5$$

and for $AD_xB_{1-x}C_2$ alloys are:

$$x_{AC} = 1/2, \quad x_{DC} = x/2 \text{ and } x_{BC} = (1-x)/2.$$

However, in the crystal structure of such alloys, the chemical bonds are arranged irregularly, as seen in Figure 1.16(a).

Quaternary alloys of ternary compounds of the second type are $ABC_{2x}D_{2(1-x)}$ alloys, shown in Figure 1.16(b). The anion sublattice contains two types of atoms, C and D. The alloys of this type are similar to the ternary alloys of binary compounds because the crystal structure of these alloys consists of the quadruples of atoms ABC_2 and ABD_2. Such quadruples can be considered as molecules of ABC_2 and ABD_2 ternary compounds with the chalcopyrite structure. However, in the crystal structure of such alloys, the quadruples of atoms are arranged irregularly,

(a)   (b)

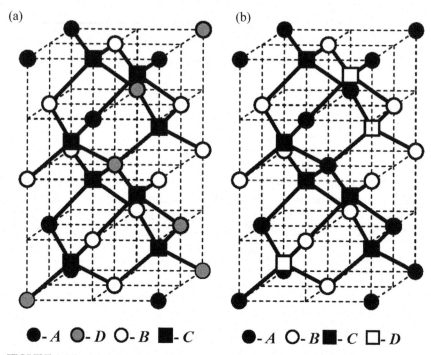

●-*A* ○-*D* ○-*B* ■-*C*   ●-*A* ○-*B* ■-*C* □-*D*

FIGURE 1.16   (a) $A_xD_{1-x}BC_2$ and (b) $ABD_{2x}C_{2(1-x)}$ alloys with the chalcopyrite structure.

as seen in Figure 1.16(b). There are four types of chemical bonds in these alloys, which are *AC*, *BC*, *AD*, and *BD*. There is a one-to-one correspondence between the elemental composition and concentrations of the chemical bonds, and the concentrations of chemical bonds are written as $x_{AC} = x_{BC} = x/2$, $x_{AD} = x_{BD} = (1 - x)/2$.

References

[1] A. Ourmazd, J.C. Bean, Observation of order-disorder transitions in strained-semiconductor systems, Phys. Rev. Lett. 55 (7) (1985) 765–768.

[2] T.S. Kuan, T.F. Kuech, W.I. Wang, E.L. Wilkie, Long-range order in III-V ternary alloys, Phys. Rev. Lett. 54 (1) (1985) 201–205.

[3] H.R. Jen, M.J. Cherng, G.B. Stringfellow, Ordered structures in GaAs$_{0.5}$Sb$_{0.5}$ alloys grown by organomerallic vapor-phase epitaxy, Appl. Phys. Lett. 48 (23) (1986) 1603–1605.

[4] H. Nakayama, H. Fugita, Direct observation of an ordered phase in a disordered In$_{1-x}$Ga$_x$As alloy, Inst. Phys. Conf. 79 (1986) 289–293.

[5] Y.E. Ihm, N. Otsuka, J. Klem, H. Morcoç, Ordering in GaAs$_{1-x}$Sb$_x$ grown by molecular-beam epitaxy, Appl. Phys. Lett. 51 (24) (1987) 2013–2015.

[6] O. Ueda, M. Takikawa, J. Komeno, I. Umebu, Atomic structure of ordered InGaP crystals grown on (001) GaAs substrates by metalorganic chemical vapor deposition, Jpn. J. Appl. Phys. 26 (11) (1987) L1824–L1827.

[7] A.G. Norman, R.E. Mallard, I.J. Murgatroyd, G.R. Booker, A.H. Moore, M.D. Scott, TED, TEM and HREM studies of atomic ordering in $Al_xIn_{1-x}As$ (x approximately 0.5) epitaxial layers, Inst. Phys. Conf. 87 (1987) 77–80.

[8] D. Korakakis, K.F. Ludwig Jr., T.D. Moustakas, Long range order in $Al_xGa_{1-x}N$ films grown by molecular beam epitaxy, Appl. Phys. Lett. 71 (1) (1997) 72–74.

[9] P. Ruterana, G. Nouet, W. Van der Stricht, I. Moerman, L. Considine, Chemical ordering in wurtzite $In_xGa_{1-x}N$ layers grown on (0001) sapphire by metalorganic vapour phase epitaxy, Appl. Phys. Lett. 72 (14) (1998) 1742–1744.

[10] K. Kasukabe, T. Yamazaki, K. Kuramochi, T. Furuzuki, I. Hashimoto, S. Ando, K. Ohkawa, Direct observation of an ordered phase in ($11\bar{2}0$) plane InGaN alloy, Jpn. J. Appl. Phys. 47 (12) (2008) 8783–8786.

Elements of Thermodynamics and Statistical Physics

In this chapter, the basic elements of equilibrium thermodynamics and equilibrium statistical physics used in the rest of this book are briefly treated. This chapter does not pretend to be a complete description of the concepts, principles, and applications of thermodynamics and statistical physics. Here there are only some of the most important basic elements necessary for the research of the thermodynamic quantities and characteristics of crystalline semiconductor alloys. The detailed consideration of the principles and applications of thermodynamics can be found in the excellent monograph by Kondepudi and Prigogine [1]. The complete exposition of the principles and applications of statistical physics is in the very good monograph by Hill [2,3]. Moreover, there are a number of textbooks in which the principles and applications of statistical physics are treated in detail [4–8].

2.1 CONCEPTS OF THERMODYNAMICS

Classical thermodynamics is based on the concepts formulated by using experimental results. Some of these concepts are listed in the next sections.

2.1.1 Mathematical Formalism of Thermodynamics

The specific formalism using the first-order homogeneous functions is the mathematical formalism of equilibrium thermodynamics.

2.1.2 Thermodynamic Systems

A system containing a very large number of particles, also called a macroscopic system, is a thermodynamic system if its characteristics can be treated by the mathematical formalism of thermodynamics. A thermodynamic system occupies a space enclosed by the boundaries that separate it from the surroundings. There are isolated, closed, and open thermodynamic systems. The system is called isolated if it does not have thermal, mechanical, and material contacts with the surroundings. The closed system can exchange heat with the surroundings and can change the volume, but does not have material contact with the surroundings. The open system can exchange matter with the surroundings. Moreover, the open system may have mechanical, thermal, or both of these contacts with the surroundings. The set of measurable parameters such as volume, pressure, temperature, numbers of particles, etc. characterizes the thermodynamic system.

2.1.3 Phase

A thermodynamic system can consist of one or more macroscopic objects named phases. A macroscopic object can be a domain or a set of domains with the same composition and structure. A one-phase system is named a homogeneous system. The systems containing two- and more phases are called heterogeneous systems.

2.1.4 Thermodynamic Variables

The thermodynamic variables are determined by the set of pairs of conjugate intensive and extensive variables. The extensive variables depend on the numbers of particles (size if the thermodynamic systems consist of condensed matter phases) in the thermodynamic system, in contrast to the intensive variables.

2.1.5 Thermodynamic Potential

The product of one pair or a sum of the several products of the conjugate thermodynamic variables are called the thermodynamic potentials attaining the absolute or relative (local) extremum (in the majority of the cases attaining the absolute or relative (local) minimum) when the thermodynamic system is in the stable or metastable state, respectively. All thermodynamic potentials are first-order homogeneous functions.

2.1.6 Fluctuations

Spontaneous perturbations of the characteristics and parameters of a thermodynamic system are called fluctuations. Normally, fluctuations are very small.

2.1.7 Thermodynamic Stability

A thermodynamic system can be in the stable, metastable, and unstable states. A thermodynamic system is in the stable state if the fluctuations cannot change its thermodynamic state. A thermodynamic system is in the metastable state if the finite fluctuations can change its thermodynamic state. A thermodynamic system is in the unstable state if the infinitesimal (negligibly small) fluctuations can lead to the transformation of its thermodynamic state.

2.2 MATHEMATICAL FORMALISM OF THERMODYNAMICS

The mathematical formalism of equilibrium thermodynamics is based on the Legendre transformations. The Legendre transformations change the roles between the dependent and independent variables of the thermodynamic potentials. These variables are pairs of the conjugate extensive-intensive quantities. All thermodynamic potentials are first-order homogeneous real functions. The first-order homogeneous function $f_\alpha(x_1,...,x_m)$ of the real independent extensive variables $x_1,...,x_m$ satisfies the condition:

$$f_\alpha(\lambda x_1, ..., \lambda x_m) = \lambda f_\alpha(x_1, ..., x_m),$$

where λ is a constant. In accordance with the Euler theorem for homogeneous functions, the following equality:

$$\frac{\partial f_\alpha(x_1, ..., x_m)}{\partial x_1} x_1 + ... + \frac{\partial f_\alpha(x_1, ..., x_m)}{\partial x_m} x_m = f_\alpha(x_1, ..., x_m)$$

is fulfilled for the first-order homogeneous function $f_\alpha(x_1,...,x_m)$. The first partial derivatives of function f_α with respect to variables x_j ($j = 1,...,m$) do not depend on λ since the equalities:

$$\frac{\partial f_\alpha(\lambda x_1, ..., \lambda x_m)}{\partial x_j} = \lambda \frac{\partial f_\alpha(\lambda x_1, ..., \lambda x_m)}{\partial x_j} = \lambda \frac{\partial f_\alpha(x_1, ..., x_m)}{\partial x_j}$$

are realized. Accordingly, the equality:

$$\frac{\partial f_\alpha(x_1, ..., x_m)}{\partial x_1} x_1 + ... + \frac{\partial f_\alpha(x_1, ..., x_m)}{\partial x_m} x_m = f_\alpha(x_1, ..., x_m), \qquad (2.2.1)$$

is fulfilled for the first-order homogeneous function $f_\alpha(x_1,...,x_m)$. Thus, if function $f_\alpha(x_1,...,x_m)$ is a thermodynamic potential, then quantities $\frac{\partial f_\alpha}{\partial x_j}$ are its intensive thermodynamic variables conjugated to extensive variables x_j. The first total differential of the function $f_\alpha(x_1,...,x_m)$ due to Eqn (2.2.1) is:

$$df_\alpha = \sum_{j=1}^{m} \frac{\partial f_\alpha}{\partial x_j} dx_j = \sum_{j=1}^{m} d\left(\frac{\partial f_\alpha}{\partial x_j} x_j\right) = \sum_{j=1}^{m} \frac{\partial f_\alpha}{\partial x_j} dx_j + \sum_{j=1}^{m} x_j d\left(\frac{\partial f_\alpha}{\partial x_j}\right),$$

where:

$$\sum_{j=1}^{m} x_j d\left(\frac{\partial f_\alpha}{\partial x_j}\right) = 0$$

which is the Gibbs–Duhem equation in equilibrium thermodynamics. The Gibbs–Duhem equation is the unique constraint and, moreover, the constraint on all intensive thermodynamic variables. Thus, quantities $\frac{\partial f_\alpha}{\partial x_j}$ are the reciprocally dependent variables. In accordance with the reasons described above, any incomplete set of the intensive thermodynamic variables is the set of the independent variables. Equation:

$$\sum_{j=1}^{m} x_j d\left(\frac{\partial f_\alpha}{\partial x_j}\right) = 0$$

can be rewritten as:

$$\sum_{j=1}^{m} x_j d\left(\frac{\partial f_\alpha}{\partial x_j}\right) = \sum_{j=1}^{m} \sum_{i=1}^{m} \frac{\partial^2 f_\alpha}{\partial x_i \partial x_j} x_j dx_i = 0.$$

If $\sum_{j=1}^{m} \sum_{i=1}^{m} \frac{\partial^2 f_\alpha}{\partial x_i \partial x_j} x_j dx_i = 0$, then the following condition:

$$\begin{vmatrix} \dfrac{\partial^2 f_\alpha}{\partial x_1^2} & \cdots & \dfrac{\partial^2 f_\alpha}{\partial x_1 \partial x_m} \\ \cdots & \cdots & \cdots \\ \dfrac{\partial^2 f_\alpha}{\partial x_m \partial x_1} & \cdots & \dfrac{\partial^2 f_\alpha}{\partial x_m^2} \end{vmatrix} = 0 \qquad (2.2.2)$$

is fulfilled. The determinant (2.2.2) is the Jacobian given by:

$$\begin{vmatrix} \dfrac{\partial}{\partial x_1} \dfrac{\partial f_\alpha}{\partial x_1} & \cdots & \dfrac{\partial}{\partial x_m} \dfrac{\partial f_\alpha}{\partial x_1} \\ \cdots & \cdots & \cdots \\ \dfrac{\partial}{\partial x_1} \dfrac{\partial f_\alpha}{\partial x_m} & \cdots & \dfrac{\partial}{\partial x_m} \dfrac{\partial f_\alpha}{\partial x_m} \end{vmatrix}. \qquad (2.2.3)$$

As the Jacobian Eqn (2.2.3) is equal to zero, a one-to-one correspondence is absent between the conjugate extensive and intensive variables x_i and $\frac{\partial f_\alpha}{\partial x_i}$. Since the Gibbs–Duhem equation is the unique constraint on the intensive variables, there is a one-to-one correspondence between any conjugate extensive x_i and intensive $\frac{\partial f_\alpha}{\partial x_i}$ variables if one of the extensive variables is fixed. In other words, any determinant:

$$
\begin{vmatrix}
\dfrac{\partial^2 f_\alpha}{\partial x_1^2} & \cdots & \dfrac{\partial^2 f_\alpha}{\partial x_1 \partial x_n} \\
\cdots & \cdots & \cdots \\
\dfrac{\partial^2 f_\alpha}{\partial x_n \partial x_1} & \cdots & \dfrac{\partial^2 f_\alpha}{\partial x_n^2}
\end{vmatrix},
\tag{2.2.4}
$$

is not equal to zero, where $x_1, \ldots, x_n$ are any n extensive variables ($n = m - 1$). The thermodynamic potential f_α should attain the absolute or relative minimum at the stable or metastable states, respectively. The minimum condition for thermodynamic potential f_α by using the Taylor's expansion is:

$$
\delta f_\alpha(x_i) = f_\alpha(x_i + \delta x_i) - f_\alpha(x_i) \approx \frac{1}{2} \sum_{i=1}^{m-1} \sum_{j=1}^{m-1} \frac{\partial^2 f_\alpha}{\partial x_i \partial x_j} \delta x_i \delta x_j > 0.
\tag{2.2.5}
$$

In accordance with the Sylvester's criterion [9], the quadratic form $\frac{1}{2} \sum_{i=1}^{m-1} \sum_{j=1}^{m-1} \frac{\partial^2 f_\alpha}{\partial x_i \partial x_j} \delta x_i \delta x_j$ is positive if the determinant (2.2.4) and all its principal minors are positive.

The thermodynamic potential corresponding to the function $f_\alpha(x_1, \ldots, x_m)$ attaining the absolute minimum in the stable state and relative minimum in the metastable state is the internal energy. For the sake of brevity, it is supposed that there is only one phase in the thermodynamic system, and the external electric and magnetic fields as well as external strain fields are assumed to be absent. Moreover, chemical reactions are also supposed absent in the system. Therefore, the entropy S, volume V, and numbers N_k of the k-th type particles are the extensive thermodynamic variables. The absolute temperature T, the negative pressure $-p$, and chemical potentials μ_k of the k-th type particles are their conjugate intensive variables, respectively. The internal energy in such a case is given by $U = TS + (-p)V + \sum_k \mu_k N_k$, which is the first-order homogeneous function. The first total differential of the internal energy is:

$$
dU = TdS + (-p)dV + \sum_k \mu_k N_k.
$$

The equation:

$$
\sum_{i=1}^{m} x_i d \left[\frac{\partial f_\alpha(x_1, \ldots, x_m)}{\partial x_i} \right] = 0
$$

is the Gibbs−Duhem equation written as $SdT + Vd(-p) + \sum_k N_k d\mu_k = 0$.

Then let us consider a new function having the same dimensionality as the function $f_\alpha(x_1,...,x_m)$ and written as:

$$f_\beta = f_\alpha - \frac{\partial f_\alpha}{\partial x_1} x_1.$$

Its first total differential is:

$$df_\beta = df_\alpha - \frac{\partial f_\alpha}{\partial x_1} dx_1 - x_1 d\left(\frac{\partial f_\alpha}{\partial x_1}\right) = -x_1 d\left(\frac{\partial f_\alpha}{\partial x_1}\right) + \sum_{i=2}^{m} \frac{\partial f_\alpha}{\partial x_i} dx_i.$$

Thus, the function $f_\beta = f_\beta\left(\frac{\partial f_\alpha}{\partial x_1}, x_2, ..., x_m\right)$ is the function of one intensive and $m-1$ extensive variables and also is the first-order homogeneous function. There are two thermodynamic potentials corresponding to the function $f_\beta = f_\beta\left(\frac{\partial f_\alpha}{\partial x_1}, x_2, ..., x_m\right)$. They are the enthalpy $H = U - (-p)V$ and the Helmholtz free energy $F = U - TS$. Their first total differentials are given, respectively, by:

$$dH = TdS - Vd(-p) + \sum_k \mu_k dN_k,$$

$$dF = -SdT + (-p)dV + \sum_k \mu_k dN_k.$$

The third function of two intensive and $m-2$ extensive variables having the same dimensionality as functions $f_\alpha(x_1,...,x_m)$ and $f_\beta = f_\beta\left(\frac{\partial f_\alpha}{\partial x_1}, x_2, ..., x_m\right)$ is:

$$f_\gamma = f_\alpha - \frac{\partial f_\alpha}{\partial x_1} x_1 - \frac{\partial f_\alpha}{\partial x_2} x_2 = f_\beta - \frac{\partial f_\alpha}{\partial x_2} x_2.$$

The thermodynamic potential corresponding to this function is the Gibbs free energy $G = U - TS - (-p)V$, which is also the first-order homogeneous function. Its first total differential is:

$$dG = -SdT - Vd(-p) + \sum_k \mu_k dN_k.$$

The same formalism as with the other thermodynamic potential (the entropy) corresponding to the function f_α attaining the absolute maximum in the stable state can be utilized. The internal energy U, volume V, and numbers N_k of the k-th type particles are extensive thermodynamic variables. The reciprocal of the absolute temperature $\frac{1}{T}$, the ratio between the pressure and absolute temperature $\frac{p}{T}$, and the ratios between the chemical potentials of the k-th type particles and absolute temperature $\frac{\mu_k}{T}$ are their conjugate intensive variables, respectively. The entropy in such a case is:

$$S = \frac{U}{T} + \frac{p}{T} + \sum_k \frac{\mu_k}{T} N_k$$

and it is the first-order homogeneous function. The first total differential of the entropy is given by:

$$dS = \frac{1}{T} dU + \frac{p}{T} dV + \sum_k \frac{\mu_k}{T} dN_k.$$

Condition (2.2.4) in this case should be changed as follows:

$$\sum_{i=1}^{m-1} x_i d\left(\frac{\partial f_\alpha}{\partial x_i}\right) = \sum_{i=1}^{m-1} \left(\frac{\partial^2 f_\alpha}{\partial x_i^2} x_i dx_i + 2 \sum_{j \neq i} \frac{\partial^2 f_\alpha}{\partial x_i \partial x_j} x_i dx_j\right) < 0.$$

2.3 CONCEPTS AND BASIC POSTULATE OF STATISTICAL PHYSICS

2.3.1 Phase Point and Phase Space

The state of a macroscopic system consisting of N particles in each instant of time can be specified by a phase point that is a set of the coordinates and momenta of all particles in a phase space consisting of mN coordinates and mN momenta for the m-dimensional ($m = 1, 2, 3$) space. The one-dimensional space can be used to consider atomic or molecular beams in vacuum. The two-dimensional space is suitable to describe the surface and physisorbed layers. The three-dimensional space is a conventional space.

2.3.2 Statistical Equilibrium

A macroscopic system is in the statistical equilibrium if the density of the phase points in the phase space does not change with time.

2.3.3 Ensemble

The principal concept of statistical physics is a concept of an ensemble. The ensemble is a set of the same macroscopic systems being in all possible states under the given conditions. These conditions can include both the external conditions such as volume and temperature, and internal characteristics of the system such as numbers of particles of each kind. The type of the ensemble is determined by the contacts with its surroundings. These contacts determine the conditions (a set of the independent variables) under which the macroscopic systems are in the ensemble. At each instant of time, a state of an ensemble is specified by a set of the phase points in the phase space. The density of the phase points in the statistical equilibrium is given by the distribution function $\rho(p,q)$, where p and q are mN momenta and mN coordinates, respectively. The ensemble average of a quantity $f(p,q)$ in the statistical equilibrium is determined by the distribution function $\rho(p,q)$ as:

$$\bar{f} = \int\int f(p,q)\rho(p,q)dpdq.$$

In the cases of the consideration of lattice systems such as crystalline semiconductor alloys, the volume is normally considered as "discrete." Accordingly, summation is used instead of integration to derive the ensemble averages.

2.3.4 Partition Function

Thermodynamic quantities and characteristics of an ensemble can be calculated from its partition function. The particle function is the function of the conditions determined by the contacts of the ensemble of the macroscopic systems with the surroundings.

2.3.5 Basic Postulate of Equilibrium Statistical Physics

A quantity f of a macroscopic system over a period of time τ is determined as a time average (the most probable value):

$$f^{TA} = \frac{1}{\tau}\int_{t}^{t+\tau} f(t)dt,$$

where $f(t)$ is the quantity f at instant of time t. The basic postulate of statistical physics is traditionally formulated for the ensembles of macroscopic systems isolated from their surroundings. This postulate is given by the time averages of quantities equal to the corresponding ensemble averages.

2.4 MICROCANONICAL ENSEMBLE

A microcanonical ensemble corresponds to a set of macroscopic systems for which the internal energy U, the volume V, and the numbers of particles of each type N_i are given conditions (given values) or, in other words, they are the independent variables. Pressure is a fluctuated quantity of such ensemble. Accordingly, the microcanonical ensemble represents the set of the isolated macroscopic systems (systems without any contacts with the surroundings). The partition function of the microcanonical ensemble is a function of the extensive variables U, V, and N_i given by:

$$\Omega(U, V, N_i) = g(U, V, N_i)$$

where $g(U,V,N_i)$ is the degeneracy factor (the number of states). The entropy of such ensemble is:

$$S = k_B \ln \Omega.$$

For crystalline alloys, the degeneracy factor is the number of arrangements of atoms over the lattice sites, and the entropy is the configurational entropy. The other thermodynamic quantities of the microcanonical ensemble are pressure and chemical potentials of the i-th type particles written, respectively, as:

$$p = k_B T \left(\frac{\partial \ln \Omega}{\partial V}\right)_{U,N_i}, \quad \mu_i = -k_B T \left(\frac{\partial \ln \Omega}{\partial N_i}\right)_{U,V}$$

The thermodynamics potential of the microcanonical ensemble and its first total differential are obtained, respectively, as:

$$-k_B T \ln \Omega = -TS = U + pV - \sum_i \mu_i N_i = H - \sum_i \mu_i N_i,$$

$$d(-TS) = dU + pdV - \sum_i \mu_i dN_i,$$

where $H = U + pV$ is the enthalpy.

Compressibility of crystalline semiconductor alloys is very small. Therefore, it can be considered in most cases (except alloys at high pressure) that the volume of such alloys does not depend on pressure. In accordance with Vegard's law, the lattice parameters of $A_x B_{1-x}$ binary semiconductor alloys and $A_x B_{1-x} C$ ternary alloys of two binary compounds are linear functions of the concentrations of the constituents. Moreover, it is normally supposed that the lattice parameter of a multicomponent semiconductor alloy is:

$$a = \sum_{i=1}^{n} a_i x_i,$$

where a_i and x_i are the lattice parameter and concentration of i-th constituent of the alloy. Hence, the volume of crystalline semiconductor alloys is determined by the numbers of particles and temperature.

For $A_xB_{1-x}C_yD_{1-y}$ quaternary alloys of four binary compounds quantities x_i are assumed as concentrations of AC, AD, BC, and BD bonds. In addition, the thermal expansion coefficients of crystalline semiconductors are very small, and therefore as a rule, the lattice parameters of semi-conductors measured at room temperature can be used in most cases. Accordingly, the partition function of the microcanonical ensemble rep-resenting a crystalline semiconductor alloy can be expressed as a function of the extensive variables U and N_i:

$$\Omega(U, N_i) = g(U, N_i).$$

In fact, this assumption is normally extended to condensed matter, i.e., solids and liquids.

2.5 CANONICAL ENSEMBLE

The absolute temperature T, the volume V, and the numbers N_i of the i-th type particles are independent variables for the canonical ensemble. Thus, the canonical ensemble is a set of the closed macroscopic systems having thermal contact with the surroundings that can be considered as an infinite thermostat with the absolute temperature T. Changes of the volume and the numbers of particles are prohibited, but the internal en-ergy and the pressure are the fluctuated quantities. The partition function of the canonical ensemble as a function of one intensive variable T and extensive variables V and N_i is written as:

$$Q(T, V, N_i) = \sum_j g_j(V, N_i) \exp\left\{-\frac{U_j}{k_B T}\right\}$$

where $g_j(V, N_i)$ is the degeneracy factor or the number of states that have the same value of the internal energy U_j. The summation extends to all possible values of the internal energy.

The thermodynamic potential of the canonical ensemble is:

$$-k_B T \ln Q = F = U - TS,$$

where F is the Helmholtz free energy and its first total differential is:

$$dF = pdV - SdT + \sum_i \mu_i dN_i.$$

The other thermodynamic quantities of the canonical ensemble such as the pressure, the entropy, and the chemical potentials of the i-th type of particles are given, respectively, by:

$$p = k_B T \left(\frac{\partial \ln Q}{\partial V}\right)_{T, N_i},$$

$$S = k_B \ln Q + k_B T \left(\frac{\partial \ln Q}{\partial T} \right)_{V,N_i},$$

$$\mu_i = -k_B T \left(\frac{\partial \ln Q}{\partial N_i} \right)_{V,T}.$$

Any macroscopic system can be represented as a closed system, and thus may be described as a canonical ensemble if the absolute temperature, volume, and numbers of particles are given values. As shown in Section 2.4, in the vast majority of descriptions of semiconductor alloys, the volume is the function of the quantities of constituents (atoms or molecules) and temperature. Thus, the partition function of the canonical ensemble can be represented as the function of the intensive variable T and extensive variables N_i:

$$Q(T, N_i) = \sum_j g_j(N_i) \exp \left\{ -\frac{U_j}{k_B T} \right\}.$$

The canonical ensemble is the most significant for the consideration of crystalline semiconductor alloys, since any such alloy can be represented as a closed system for which the temperature is a given value. Therefore, the canonical ensemble is mainly used in this book.

2.6 ISOTHERMAL–ISOBARIC ENSEMBLE

The absolute temperature T, the pressure p, and the numbers N_i of the i-th type particles are the independent variables of the isothermal–isobaric ensemble that can be represented as a set of the closed macroscopic systems having mechanical and thermal contacts with the surroundings. The internal energy and the volume are the fluctuated quantities of such an ensemble. The partition function of the ensemble for isothermal–isobaric macroscopic systems in the discrete form more suitable for lattice systems (crystalline semiconductor alloys) considered in this book is written as:

$$\Delta(T, p, N_i) = \sum_j \sum_l g_{jl}(p, N_i) \exp \left\{ -\frac{U_j + pV_l}{k_B T} \right\},$$

where $g_{jl}(p,N_i)$ is the degeneracy factor or the number of states that have the same value of the internal energy U_j and volume V_l. The summation is taken over all possible values of the internal energy and volume.

The thermodynamic potential of the isothermal–isobaric ensemble is:

$$-k_B T \ln \Delta = G = F + pV = U - TS + pV = H - TS = \sum_i \mu_i N_i,$$

where G is the Gibbs free energy and its first total differential is:

$$dG = -Vdp + SdT + \sum_i \mu_i dN_i,$$

The other thermodynamic quantities of the isothermal–isobaric ensemble such as the entropy, the volume, and the chemical potentials of the i-th type of particles are given, respectively, by:

$$V = -k_B T \left(\frac{\partial \ln \Delta}{\partial p} \right)_{T,N_i},$$

$$S = k_B \ln \Delta + k_B T \left(\frac{\partial \ln \Delta}{\partial T} \right)_{p,N_i},$$

$$\mu_i = -k_B T \left(\frac{\partial \ln \Delta}{\partial N_i} \right)_{T,p}.$$

For crystalline semiconductor alloys, as explained in Section 2.4, the volume of the ensemble may be also determined by the temperature and numbers of particles.

2.7 GRAND CANONICAL ENSEMBLE

The absolute temperature T, the volume V, and the chemical potentials μ_i of particles of each i-th type are the independent variables for the grand canonical ensemble, which can be represented as a set of open macroscopic systems with fixed walls permeable for heat and particles. The internal energy, the pressure, and the numbers of particles are the fluctuated quantities for it. The partition function of the grand canonical ensemble is:

$$\Xi(T, V, \mu_i) = \sum_{j,i,N_i} g_{jiN_i}(V) \exp\left\{ -\frac{U_j - \mu_i N_i}{k_B T} \right\},$$

where $g_{jiN_i}(V)$ is the degeneracy factor or the number of states that have the same value of the internal energy U_j and the numbers of particles N_i of each i-th type. The summation is produced for all possible values of the internal energy U_j and all possible numbers of particles of each type.

The thermodynamic potential of the grand canonical ensemble is obtained as:

$$-k_B T \ln \Xi = -pV = U - TS - \sum_{i,N_i} \mu_i N_i$$

and its total differential is:

$$d(-pV) = -pdV - SdT - \sum_{i,N_i} N_i d\mu_i.$$

The other thermodynamic quantities of the grand canonical ensemble such as the pressure, the entropy, and the numbers of particles of different types are given, respectively, by:

$$p = k_B T \left(\frac{\partial \ln \Xi}{\partial V} \right)_{T,\mu_i},$$

$$S = k_B \ln \Xi + k_B \left(\frac{\partial \ln \Xi}{\partial T} \right)_{V,\mu_i},$$

$$N_i = k_B T \left(\frac{\partial \ln \Xi}{\partial \mu_i} \right)_{T,V,\mu_{j \neq i}}.$$

The consideration of semiconductor alloys as grand canonical ensembles may be difficult or inconvenient (e.g., multicomponent alloys of semiconductor compounds with the crystal lattices consisting of two or more mixed sublattices). Therefore, the grand canonical ensemble is not used in this book.

2.8 MACROSCOPIC SYSTEMS

The mathematical formalism of classical thermodynamics described in Section 2.2 contains the fundamental limitation for the functions that can be treated as thermodynamic quantities. This restriction results in the use of the first-order homogeneous functions as the thermodynamic potentials. It imposes the limit on the minimal size of a system that can be considered by using the methods of classical thermodynamics. Statistical physics also normally deals with the systems containing very large numbers of particles. A system with a size larger than the minimal size in accordance with the restriction introduced by the mathematical formalism is denoted macroscopic. The systems that do not satisfy this criterion can be considered by the approaches described in the book devoted to thermodynamics of small systems [10]. Moreover, classical thermodynamics as well as statistical physics (except for a few special cases) deal with the systems in which the fluctuations should be negligibly small and that also depend on the size of the system. These limitations permit estimating the minimal size of the macroscopic system. Some criteria of macroscopicity of a system are given next.

2.8.1 Additivity of the Configurational Entropy

Any thermodynamic system may be characterized among other quantities by its configurational entropy. This entropy should be the first-order homogeneous function and, thus, an additive function. The additivity condition of the configurational entropy as a function of numbers of particles is discussed here. Let us consider the random binary mixture consisting of particles A and B. The configurational entropy of the mixture may be represented as the entropy of the system that is a unique object and the entropy of the system consisting of the same numbers of particles A and B that is a set of two parts.

The configurational entropy of the system that is a unique object is:

$$S_1 = k_B \ln \frac{(N_A + N_B)!}{N_A! N_B!}.$$

After using Stirling's formula $N! \approx N^N e^{-N} \sqrt{2\pi N}$, the entropy S_1 can be rewritten as:

$$S_1 = -k_B N_A \ln \frac{N_A}{N_A + N_B} - k_B N_B \ln \frac{N_B}{N_A + N_B} + k_B \ln \sqrt{2\pi(N_A + N_B)}$$
$$- k_B \ln \sqrt{2\pi N_A} - k_B \ln \sqrt{2\pi N_B}.$$

The configurational entropy of the system consisting of two parts is:

$$S_2 = k_B \ln \frac{\left[\frac{m(N_A+N_B)}{m+n}\right]!}{\left(\frac{mN_A}{m+n}\right)! \left(\frac{mN_B}{m+n}\right)!} + k_B \ln \frac{\left[\frac{n(N_A+N_B)}{m+n}\right]!}{\left(\frac{nN_A}{m+n}\right)! \left(\frac{nN_B}{m+n}\right)!}$$

where $\frac{m}{m+n}$ and $\frac{n}{m+n}$ are the portions of particles A (B) in the first and second parts of the system, respectively, and $m + n = 1$. The difference between the entropies S_1 and S_2, by using Stirling's formula, is written as:

$$S_1 - S_2 = -k_B \ln \sqrt{2\pi(N_A + N_B)} + k_B \ln \sqrt{2\pi N_A} + k_B \ln \sqrt{2\pi N_B}$$
$$+ k_B \ln \sqrt{mn}.$$

The additivity condition for the configurational entropy is fulfilled if the ratio $\frac{S_1 - S_2}{S_1}$ is negligibly small. If $N_A = N_B$ and $m = n$ and if the number of particles A is larger than 10^7, then the value of the ratio $\frac{S_1 - S_2}{S_1}$ should be less than 10^{-6}.

2.8.2 Relative Fluctuation of the Internal Energy in the Crystalline Alloy

Let us consider the relative fluctuation of the internal energy in binary crystalline alloy (lattice system) that is represented as the canonical ensemble. The relative fluctuation of the internal energy in the canonical ensemble is written as $\sqrt{(U - \overline{U})^2}$, where $\overline{U}$ is the ensemble average of the internal energy that is given for a lattice system by:

$$\overline{U} = \frac{\sum_j g_j(V, N_i) U_j \exp\left\{-\frac{U_j}{k_B T}\right\}}{\sum_j g_j(V, N_i) \exp\left\{-\frac{U_j}{k_B T}\right\}},$$

where $g_j(V,N_i)$ is the degeneracy factor or the number of states that have the same value of the internal energy U_j and at the given volume and at the given numbers of particles. The canonical ensemble average of the internal energy can be presented also as $\overline{U} = F - \frac{1}{T}\left(\frac{\partial F}{\partial T}\right)$, where F is the Helmholtz free energy. The definition of the partition function of the canonical ensemble:

$$Q = \sum_j g_j(V, N_i) \exp\left\{-\frac{U_j}{k_B T}\right\} = \exp\left\{-\frac{F}{k_B T}\right\}$$

results in the expression $\sum_j g_j(V, N_i) \exp\left\{-\frac{U_j - F}{k_B T}\right\} = 1$. The differentiation of this expression with respect to the absolute temperature gives us:

$$\frac{\partial}{\partial T} \sum_j g_j \exp\left\{-\frac{U_j - F}{k_B T}\right\}$$

$$= \frac{1}{k_B T^2} \sum_j g_j \left[U_j - F + T\left(\frac{\partial F}{\partial T}\right)\right] \exp\left\{-\frac{U_j - F}{k_B T}\right\}$$

$$= \frac{1}{k_B T^2} \sum_j g_j \left(U_j - \overline{U}\right) \exp\left\{-\frac{U_j - F}{k_B T}\right\} - 0.$$

The differentiation of expression $\sum_j g_j(U_j - \overline{U}) \exp\left\{-\frac{U_j - F}{k_B T}\right\}$ with respect to the absolute temperature results in:

$$\frac{\partial}{\partial T} \sum_j g_j \left(U_j - \overline{U}\right) \exp\left\{-\frac{U_j - F}{k_B T}\right\}$$

$$= \sum_j g_j \left(U_j - \overline{U}\right) \frac{\partial}{\partial T} \exp\left\{-\frac{U_j - F}{k_B T}\right\} - \frac{\partial \overline{U}}{\partial T} \sum_j g_j \exp\left\{-\frac{U_j - F}{k_B T}\right\} = 0$$

which can be rewritten as:

$$\frac{\partial \overline{U}}{\partial T} - \frac{1}{k_B T^2} \frac{\sum_j g_j \left(U_j - \overline{U}\right)^2 \exp\left\{-\frac{U_j}{k_B T}\right\}}{\sum_j g_j \exp\left\{-\frac{U_j}{k_B T}\right\}} = \frac{\partial \overline{U}}{\partial T} - \frac{1}{k_B T^2} \overline{(\overline{U} - U)^2} = 0.$$

Thus, the mean square fluctuation is:

$$\overline{(\overline{U} - U)^2} = k_B T^2 \frac{\partial \overline{U}}{\partial T} = k_B T^2 C_p.$$

For sufficiently high temperatures ($T > 0\,^\circ$C) the specific heat of a single crystalline solid depends, in fact, only on the number of atoms. In particular, the specific heat per atom should be close for a number of different crystalline solids and equal to $3k_B$ in accordance with the Dulong and Petit's law. At conventional temperatures (e.g., room temperature), the Dulong and Petit's law is well satisfied for many crystalline solids, for example, for crystalline Ge and Si. Accordingly, the obtained results can be applied to estimate the relative fluctuations of the internal energy in Ge_xSi_{1-x} alloys. Let N be a total number of Ge and Si atoms in the system. In such a case, the ensemble average of the internal energy and relative fluctuation of the internal energy are:

$$\overline{U} = 3k_B TN, \sqrt{\overline{(U - \overline{U})^2}} = \sqrt{3N} k_B T.$$

The relative fluctuation of the internal energy and the ensemble average of the internal energy should be negligibly small for a thermodynamic system. If the ratio between the relative fluctuation and the ensemble average is:

$$\frac{\sqrt{\overline{(U - \overline{U})^2}}}{\overline{U}} = \sqrt{\frac{3}{N}} \leq 10^{-6}.$$

then the total number of Ge and Si atoms in the system should be more than 3×10^{12}.

2.8.3 Additivity of the Internal Energy

The internal energy of the thermodynamic system is an additive function of the total number of atoms. However, the contributions of the surface atoms and atoms in bulk of the same crystalline solid are normally different. Therefore, the number of the surface atoms should be negligibly small in comparison with the number of atoms in bulk to accomplish the additivity condition of the internal energy. The number of the surface atoms is not directly proportional to the total number of atoms. Indeed, the separation of the system into several parts leads to an increase of the number of surface atoms and a decrease of the number of atoms in bulk. Hence, the contribution of the surface atoms to the internal energy of the system should be negligibly small in comparison with that in bulk. In such a case, the internal energy of a macroscopic system will be additive.

The ratio between the numbers of surface atoms and atoms in bulk in a three-dimensional macroscopic system (e.g., bulk crystal) can be estimated as follows. Let N be the total number of atoms. The number of the surface atoms in a crystalline solid of the cubic shape is equal to $6\sqrt[3]{N^2}$, where N is the total number of atoms. The ratio between the numbers of surface atoms and atoms in the bulk is:

$$\frac{N^S}{N^B} = \frac{6\sqrt[3]{N^2}}{N - 6\sqrt[3]{N^2}}.$$

This ratio should be negligibly small. The total number of atoms in the solid should be not less than $N = 10^{20}$ to obtain the value of the ratio $\frac{N^S}{N^B} \approx 10^{-6}$.

The ratio between the numbers of surface and other atoms in a two-dimensional macroscopic system (e.g., physisorbed layer) is obtained as follows. Surface atoms in the two-dimensional system are atoms situated on the boundary of the macroscopic system forming a one-dimensional system (line). Let N be the total number of atoms forming the square-shaped area. The number of atoms on the boundary is:

$$N^B = 4\sqrt{N}.$$

The ratio between the numbers of atoms on the boundary and other atoms expressed as:

$$\frac{4\sqrt{N}}{N - 4\sqrt{N}} = \frac{4}{\sqrt{N} - 4}$$

should be negligibly small. The total number of atoms has to be not less than $N = 10^{13}$ to obtain the value of the ratio $\approx 10^{-6}$.

The ratio between the numbers of boundary and other atoms in a one-dimensional macroscopic system (chain) can be estimated as follows. Let N be the total number of atoms. There are two boundary atoms in the open-boundary chain. The ratio between the number of boundary and other atoms is:

$$\frac{2}{N - 2}.$$

The total number of atoms in the solid should be not less than $N = 10^6$ in order to get the value of the ratio $\approx 10^{-6}$.

The estimates demonstrate that additivity of the internal energy should be fulfilled for the same linear sizes of the three-, two- and one-dimensional systems.

2.9 FREE ENERGIES OF CONDENSED MATTER

The Helmholtz $F(T,V,N_i) = U - TS$ and Gibbs $G(T,p,N_i) = U - TS + pV$ free energies are the thermodynamic potentials of the canonical and isothermal–isobaric ensembles, respectively. As shown in Sections 2.4–2.6, the partition functions of both ensembles representing condensed matter (solids and liquids) under pressure on the order of 1 atm and less may be considered as the functions of the absolute temperature and the numbers of particles only. This occurs due to the significantly higher density of condensed matter in comparison with the density of gases under such pressure. In fact, the density of solids and liquids normally is more than 10^3 times higher than the density of the same substances in the gaseous state at atmospheric pressure. Accordingly, the term pV, which is the difference between the Gibbs and Helmholtz free energies, is more than 1000 times less for condensed matter. For Ge_xSi_{1-x} alloys, the value of this term per mole at room temperature and atmospheric pressure is:

$$pv = 1.206 + 0.157x, \text{ J mol}^{-1}.$$

The values of this term for other semiconductor alloys are also on the order of several J mol^{-1}. The variation of the term pv for Ge_xSi_{1-x} alloys from absolute zero to room temperature at atmospheric pressure is:

$$pv = pv_0 \left(1 + \int_0^{298.15} \alpha_{th} dT \right),$$

where $v_0 = v_{0,Ge}x + v_{0,Si}(1 - x)$, $v_{0,Ge}$ is the molar volume of Ge at temperature of $T = 0$ K, $\alpha_{th} = \frac{1}{v}\left(\frac{\partial v}{\partial T}\right)_p = \alpha_{th,Ge}x + \alpha_{th,Si}(1 - x)$ is the thermal expansion coefficient of the alloy, and $\alpha_{th,Ge} \approx 2.7 \times 10^{-6}$ K^{-1} and $\alpha_{th,Si} \approx 1.3 \times 10^{-6}$ K^{-1} are the thermal expansion coefficients of Ge and Si, respectively. The variation of the term pv for $Ge_{0.5}Si_{0.5}$ alloy $\Delta pv = pv - pv_0 = 7.659 \times 10^{-4}$ J mol^{-1}.

The entropy of Ge_xSi_{1-x} alloys with the random distribution of atoms is the sum of the entropies of the constituents and configurational entropy given by:

$$s = s_{Ge}x + s_{Si}(1 - x) + s^{Conf}$$
$$= s_{Ge}x + s_{Si}(1 - x) + R[x \ln x + (1 - x)\ln(1 - x)].$$

The entropies of the constituents at room temperature and atmospheric pressure are the standard entropies equal to $s_{Ge}^0 = 31.09$ J mol^{-1} K^{-1} and

$s^0_{Si} = 18.83$ J mol^{-1} K^{-1}. Thus the molar entropy of Ge$_x$Si$_{1-x}$ alloys at room temperature and atmospheric pressure is:

$$s = 18.83 + 12.26x + 8.316[x \ln x + (1 - x)\ln(1 - x)], \text{ J mol}^{-1} \text{ K}^{-1}.$$

Accordingly, the absolute value of the entropy term $-Ts$ for these alloys at standard temperature and pressure is considerably greater than the term pv, and this value for Ge$_{0.5}$Si$_{0.5}$ alloy is equal to 9.156 kJ mol^{-1}.

The internal energy of Ge$_x$Si$_{1-x}$ alloys is the sum of the internal energy of the constituents and internal strain energy. The internal strain energy in the framework of the model of regular solutions (Chapter 3) can be expressed as:

$$u^{\text{Strain}} = \alpha_{Ge-Si}x(1 - x),$$

where $\alpha_{Ge-Si} = 5.022$ kJ mol^{-1} is the interaction parameter between Ge and Si atoms [11]. In accordance with the model of regular solutions, this energy does not depend on temperature. The variations of the enthalpies of the constituents from absolute zero to room temperature are $\Delta h_{Ge} = 4.624$ kJ mol^{-1} and $\Delta h_{Si} = 3.218$ kJ mol^{-1}. Accordingly, the variation of the enthalpy of Ge$_x$Si$_{1-x}$ alloy can be estimated as:

$$\Delta h = \Delta h_{Si} + (\Delta h_{Ge} - \Delta h_{Si})x >> p\Delta V.$$

The fulfilled estimates show that in the most cases the term pV may be ignored when examining the thermodynamic quantities and characteristics of crystalline semiconductor alloys.

2.10 EQUIVALENCE OF ENSEMBLES

To demonstrate the equivalence of the ensembles, the internal energies and their fluctuations in alloys represented as the canonical, microcanonical, and isothermal–isobaric ensembles are considered. The fluctuations of pressure or volume do not have important implication for alloys since the internal energy U is much greater than term pV for condensed matter, as shown in Section 2.9.

2.10.1 The Equivalence of the Canonical and Microcanonical Ensembles

The internal energy of a system in the canonical ensemble is undetermined in an infinite interval. At the same time, the internal energy of any system in the microcanonical ensemble is a given (fixed) value.

The partition function of the canonical ensemble as a function of one intensive variable T (absolute temperature) and extensive variables V (volume) and N_i (numbers of particles) is:

$$Q(T, V, N_i) = \sum_j g_j(V, N_i) \exp\left\{-\frac{U_j}{k_B T}\right\}$$

where $g_j(V, N_i)$ is the degeneracy factor or number of states that have the same value of the internal energy U_j. As shown in Section 2.8, the relative fluctuation of the internal energy $\sqrt{(U - \overline{U})^2}$, where $\overline{U} = \dfrac{\sum\limits_j g_j(V, N_i) U_j \exp\left\{-\frac{U_j}{k_B T}\right\}}{\sum\limits_j g_j(V, N_i) \exp\left\{-\frac{U_j}{k_B T}\right\}}$

is the ensemble average of the internal energy in $Ge_x Si_{1-x}$ bulk semiconductor alloys, and is given by:

$$\sqrt{(U - \overline{U})^2} = \sqrt{3N k_B T}.$$

The ratio between the relative fluctuation of the internal energy and the ensemble average of the internal energy is:

$$\frac{\sqrt{(U - \overline{U})^2}}{\overline{U}} = \sqrt{\frac{3}{N}},$$

where $N = \sum_{i=1}^2 N_i$ is the total number of Si and Ge atoms.

A deviation from the additivity of the internal energy of bulk alloys is determined by the ratio between the numbers of surface atoms and atoms in bulk that introduce different energies in the total internal energy. To obtains the ratio between the number of the surface atoms and total number of atoms less than 10^{-6}, the total number of atoms should be more than 10^{20}. In such a case, the ratio between the relative fluctuation of the internal energy and the ensemble average of the internal energy is on the order of 10^{-10} or less. Thus, the probability distribution of the internal energy (the interval δU of the internal energies in which the probability is not negligibly small) in the canonical ensemble is very sharp. In other words, the fluctuations of the internal energy in the canonical ensemble are insignificant.

2.10.2 Equivalence of the Canonical and Isothermal–Isobaric Ensembles

The partition function of the isothermal–isobaric ensemble is written as:

$$\Delta(T, p, N_i) = \sum_j \sum_l g_{jl}(p, N_i) \exp\left\{-\frac{U_j + p V_l}{k_B T}\right\},$$

where $g_{jl}(p,N_i)$ is the degeneracy factor or number of states that have the same value of the internal energy U_j and volume V_l. The summation is taken over all possible values of the internal energy and volume. Due to the condition $\overline{U} \gg p\overline{V}$ for the condensed matter, where $\overline{U}$ and $\overline{V}$ are the ensemble averages of the internal energy and volume, the ensemble average of the internal energy can be used instead of the ensemble average of the enthalpy. After that, all considerations of the fluctuations of the internal energy in the canonical ensemble can be applied. Finally, the ratio between the relative fluctuation of the internal energy and the ensemble average of the internal energy is:

$$\frac{\sqrt{\left(U - \overline{U}\right)^2}}{\overline{U}} = \sqrt{\frac{3}{N}}.$$

where $N = \sum_{i=1}^{2} N_i$ is the total number of Si and Ge atoms in Ge_xSi_{1-x} alloy. Thus, the probability distribution of the internal energy in the iso-thermal–isobaric ensemble representing Ge_xSi_{1-x} alloy is very sharp and the fluctuations of the internal energy in the isothermal–isobaric ensemble are negligible.

2.11 SEPARATION OF DEGREES OF FREEDOM

This book is devoted to the semiconductor crystalline substitutional alloys, their thermodynamic stability, and the phase transformations in them. Therefore, such alloys are represented here as macroscopic lattice systems. To describe crystalline solids, the degrees of freedom relating to the positions of atoms are normally considered as degrees of freedom separable from all other degrees of freedom. The other degrees of freedom not dependent on the distribution of atoms over the lattice sites are considered in this book as internal degrees of freedom. The electronic (the degrees of freedom relating to electrons of the internal shells) and nuclear degrees of freedom are normally assumed to be the internal degrees of freedom separable from all other degrees of freedom [7]. In some cases in the derivation of crystalline solids, the vibrational degrees of freedom can be also represented as internal degrees of freedom. In any case, the vibrational degrees of freedom are considered as separable from the electronic and nuclear degrees of freedom. Accordingly, the partition function of a crystalline semiconductor alloy represented by the canonical ensemble is:

$$Q = Q_{Int}Q_{Conf}, \tag{2.11.1}$$

where Q_{Int} is the internal partition function relating to the internal degrees of freedom and Q_{Conf} is the configurational partition function depending on the positions of atoms over the lattice sites. As mentioned earlier, the partition function corresponding to the oscillations of atoms should be a part of the internal partition function or part of the configurational partition function. In the latter case, the degrees of freedom corresponding to the oscillations are also separable from the other degrees of freedom. The Helmholtz free energy of a crystal can, in such a case, be written as:

$$F = U - TS = -k_B T \ln Q = -k_B T \ln Q_{Int} - k_B T \ln Q_{Conf}, \qquad (2.11.2)$$

where U, T, and S are the total internal energy, the absolute temperature, and the total entropy, respectively. It follows from this formula that the part of the free energy dependent on the distribution of atoms and transformations of a crystalline alloy, such as mixing of several crystals in one and vice versa, or an occurrence of a superstructure as well as its disappearance, are determined by the configurational partition function Q_{Conf}. Accordingly, in the consideration of characteristics of the alloys and processes in them, all contributions to the Helmholtz free energy due to the internal partition function Q_{Int} can be neglected. The total internal energy of semiconductor alloys given in Eqn (2.11.2) is expressed as a sum of the energies determined by the internal degrees of freedom and degrees of freedom represented by the configurational partition function.

The concentrations of constituents in alloys of the elemental semi-conductors and in $A_xB_{1-x}C$, AB_xC_{1-x}, ternary and $A_xB_yC_{1-x-y}D$, $AB_xC_yD_{1-x-y}$ quaternary alloys (alloys with one mixed sublattice) of semiconductor compounds do not depend on the distribution of atoms over the lattice sites (see Chapter 1, Sections 1.6–1.8). Accordingly, the internal energy and entropy of the constituents of such alloys are obtained from the internal partition functions. The internal energy caused by mixing of the constituents and configurational entropy are represented by the configurational partition functions. The concentrations of chemical bonds in $A_xB_{1-x}C_yD_{1-y}$ and more *multi-atomic* semiconductor alloys (alloys with two or more mixed sublattices), as shown in Chapter 1, Sections1.9 and 1.10, are strongly controlled by the distribution of atoms over the lattice sites. Therefore, the internal partition function in Eqn (2.11.2) are represented only by the electronic (the degrees of freedom relating to electrons of the internal shells) and nuclear degrees of freedom. The internal energy due to the chemical bonds and internal energy of mixing of the constituents as well as the configurational entropy are related to the configurational partition function in Eqn (2.11.2).

It is reasonable to assume that the degrees of freedom corresponding to the oscillations of atoms in the alloys of the elemental semiconductors and alloys of binary compounds with one mixed sublattice are the internal degrees of freedom involved in the internal partition function in Eqn (2.11.2). The part of the partition function controlled by the degrees of freedom relating to the vibrations of atoms in $A_xB_{1-x}C_yD_{1-y}$ and more multi-atomic alloys with two or more mixed sublattices should be in the configurational partition function. In such alloys, the oscillations depend strongly on the distribution of atoms over the lattice sites determining the concentrations of bonds of the constituent compounds.

References

[1] D. Kondepudi, I. Prigogine, Modern Thermodynamics, John Wiley & Sons, Chichester, 1998, 1999, 2002, 2004, 2005, 2006, 2007.

[2] T.L. Hill, Statistical Mechanics, McGraw-Hill Book Company, Inc., New York, 1956.

[3] T.L. Hill, Statistical Mechanics, Dover Publications, Inc., New York, 1987.

[4] P.K. Pathria, Statistical Mechanics, second ed., Elsevier, Amsterdam, 1996, 1997, 1998, 1999, 2000, 2001, 2003, 2004, 2005 (twice).

[5] J.E. Mayer, M.G. Mayer, Statistical Mechanics, second ed., John Wiley & Sons, New York, 1977.

[6] K. Huang, Statistical Mechanics, second ed., John Wiley & Sons, New York, 1987.

[7] L.E. Reichl, A Modern Course in Statistical Mechanics, second ed., John Wiley & Sons, New York, 1997.

[8] D.A. McQuarrie, Statistical Mechanics, University Science Books, Sausalito, California, 2000.

[9] G.A. Korn, T.M. Korn, Mathematical Handbook, second ed., McGraw-Hill Book Company, New York, 1968 (Chapter 13).

[10] T.L. Hill, Thermodynamics of Small Systems, Dover Publications, Inc, Mineola, 1994.

[11] J.L. Martins, A. Zunger, Bond lengths around isovalent impurities and in semiconductor solid solutions, Phys. Rev. B 30 (10) (1984) 6217–6220.

Regular Solutions

The study of thermodynamic properties and characteristics of crystalline semiconductor alloys is carried out by various lattice models. Among them the regular solution model is the most applicable due to its relative simplicity in comparison with the other models. Any model has both advantages and disadvantages, and is usually chosen for reasons that, in considering the properties or characteristics, you can achieve the desired results. One of the very important advantages of the regular solution model is its simplicity, making it possible to use different approximations since, as a rule, the exact solutions of the problems are inaccessible. It is mostly related to the calculation of the configurational entropy of the multicomponent semiconductor alloys that can be estimated with a high accuracy by using the regular solution model. Therefore, the consideration of semiconductor alloys represented as regular solutions is a significant part of this book. The replacement of the Gibbs free energy by the Helmholtz free energy and vice versa established in Chapter 2, Section 2.9 for condensed matter is widely used in this chapter.

3.1 REGULAR SOLUTION MODEL

The lattice model of liquid mixtures with the random distribution of molecules called a regular solution model was introduced by Hildebrand [1]. Later, Guggenheim [2] extended the theory of regular solutions to the crystalline substitutional alloys and wider class of liquid solutions. The assumption on separation of the degrees of freedom is the basis of this model. That is why the regular solution model is widely used to describe the semiconductor alloys. In most cases, the absolute temperature, numbers of each type of atom, and volume of a semiconductor alloy are given values. Therefore, the formalism of the canonical ensemble is very appropriate to describe such alloys and will play a major role in the further considerations. The theory of regular solutions deriving crystalline substitutional alloys represented as canonical ensembles is based on the following assumptions:

1. Atoms are distributed over the lattice sites and each atom may occupy any lattice site.
2. The location (coordinates) of the lattice sites does not depend on the distribution of atoms and composition of regular solution.
3. The internal degrees of freedom as well as the degrees of freedom relating to the positions of atoms are considered as separable. Thus, the partition function of a regular solution is given by:

$$Q = Q_{Int} Q_{Conf},$$

where Q_{Int} and Q_{Conf} are the internal and configurational partition functions, respectively.

4. Only the configurational partition function depends on the distribution of atoms over the lattice sites. The interactions between the nearest atoms are only taken into account and the interaction energy between the nearest neighbors is independent of their surroundings.
5. The vibration motion of atoms is determined by the internal degrees of freedom of atoms.

The thermodynamic quantities of the crystalline solid consisting of atoms A and represented as a canonical ensemble are obtained from its partition function. According to the assumptions of the regular solution model described above (1–5), the partition function is

$$Q = Q_{Int} Q_{Conf} = \left(\sigma_A \exp\left\{ -\frac{\chi_A}{k_B T} \right\} \right)^{N_A}$$

where σ_A is the internal partition function of atom A, χ_A is the internal energy of atom A, and N_A is the number of atoms A. As follows from the assumptions (1–5), the internal energy of atoms A is a sum of the pair interactions between the nearest neighbors A and, therefore, the partition function can be rewritten as:

$$Q = \sigma_A^{N_A} \left(\exp\left\{ -\frac{u_{AA}}{k_B T} \right\} \right)^{N_{AA}}, \tag{3.1.1}$$

where N_{AA} is the number of pairs of the nearest neighbors, or in the other words, the number of bonds between the nearest neighbors; $u_{AA} = \frac{2\chi_A}{z_1}$ is the internal energy per bond AA; and z_1 is the coordination number of the nearest neighbors. At the given temperature and number of atoms, the internal energy of the crystalline solid is a fixed value. Thus, the partition function (3.1.1) is equivalent to the partition function of the micro-canonical ensemble. For the single crystalline solid, consisting of atoms A and B and represented as $A_x B_{1-x}$ regular solution, in accordance with the assumptions (1–5) given above, each pair of the nearest neighbor atoms A and each pair of the nearest neighbor atoms B contribute the factors $\exp\left\{-\frac{u_{AA}}{k_B T}\right\}$ and $\exp\left\{-\frac{u_{BB}}{k_B T}\right\}$, respectively, to the configurational partition function. It is further assumed that each pair of the nearest neighbor atoms A and B contributes the factor that, by analogy with the factors $\exp\left\{-\frac{u_{AA}}{k_B T}\right\}$ and $\exp\left\{-\frac{u_{BB}}{k_B T}\right\}$, is $\exp\left\{-\frac{u_{AB}}{k_B T}\right\}$, where u_{AB} is the internal energy per bond AB. If the crystalline solid is composed of N_A atoms A and N_B atoms B with a completely defined arrangement of the atoms over the lattice sites, it is considered that each lattice site is specified either as occupied by atom A or as occupied by atom B. For such specified location of atoms, the partition function of the crystalline solid is:

$$Q = \sigma_A^{N_A} \sigma_B^{N_B} \left(\exp\left\{-\frac{u_{AA}}{k_B T}\right\}\right)^{N_{AA}} \left(\exp\left\{-\frac{u_{BB}}{k_B T}\right\}\right)^{N_{BB}} \left(\exp\left\{-\frac{u_{AB}}{k_B T}\right\}\right)^{N_{AB}},$$

(3.1.2)

where N_{AA}, N_{BB}, and N_{AB} are the numbers of the AA, BB, and AB nearest neighbor pairs, respectively. By direct calculation, the relations between the numbers of the nearest neighbor pairs and the numbers of atoms are obtained as:

$$N_{AA} = \frac{z_1 N_A - N_{AB}}{2},$$

$$N_{BB} = \frac{z_1 N_B - N_{AB}}{2}.$$

Thus, at the given temperature and given numbers of atoms, the partition function (3.1.2) is the function of the number of AB pairs only.

If the distribution of atoms on the lattice sites is not specified, the partition function is:

$$Q = \sigma_A^{N_A} \sigma_B^{N_B} \sum_{N_{AB}} g(N_A, N_B, N_{AB}) \left(\exp\left\{-\frac{u_{AA}}{k_B T}\right\}\right)^{\frac{z_1 N_A - N_{AB}}{2}}$$
$$\times \left(\exp\left\{-\frac{u_{BB}}{k_B T}\right\}\right)^{\frac{z_1 N_B - N_{AB}}{2}} \left(\exp\left\{-\frac{u_{AB}}{k_B T}\right\}\right)^{N_{AB}},$$

(3.1.3)

where $g(N_A, N_B, N_{AB})$ is the number of different configurations of atoms arranged over the lattice sites with the value of the internal energy specified by the number of AB pairs. Such partition function derives the crystalline solid alloy represented as the canonical ensemble since the numbers of atoms (volume) and temperature are fixed, but the internal energy is not a given value. The immediate use of formula (3.1.3) is inconvenient, since the thermodynamic quantities should be obtained from the natural logarithm of the partition function, but processing of a logarithm of a sum is difficult. However, due to the negligibly small fluctuations of the internal energy in a thermodynamic system represented as a canonical ensemble, the value of the sum (3.1.3) is very close to the value of the maximal item of this sum. In other words, the value of the sum (3.1.3) of all items except the maximal item is significantly smaller than the value of the maximal item. Accordingly, the partition function (3.1.3) can be rewritten as:

$$Q \approx \sigma_A^{N_A} \sigma_B^{N_B} g\left(N_A, N_B, N_{AB}^{\#}\right) \left(\exp\left\{-\frac{u_{AA}}{k_B T}\right\}\right)^{\frac{z_1 N_A - N_{AB}^{\#}}{2}}$$

$$\times \left(\exp\left\{-\frac{u_{BB}}{k_B T}\right\}\right)^{\frac{z_1 N_B - N_{AB}^{\#}}{2}} \left(\exp\left\{-\frac{u_{AB}}{k_B T}\right\}\right)^{N_{AB}^{\#}},$$

where $N_{AB}^{\#}$ is the number of AB bonds corresponding to the maximal item in the sum (3.1.3).

The Helmholtz free energy of $A_x B_{1-x}$ regular solution with the specified number of $N_{AB}^{\#}$ is:

$$F = -k_B T \ln Q = -N_A k_B T \ln \sigma_A - N_B k_B T \ln \sigma_B + \chi_A N_A + \chi_B N_B$$

$$+ \left(u_{AB} - \frac{u_{AA} + u_{BB}}{2}\right) N_{AB}^{\#} - k_B T \ln g\left(N, N_A, N_{AB}^{\#}\right).$$

$$\text{or } F = F_A + F_B + U^M - TS^M,$$

where $F_A = -N_A k_B T \ln \sigma_A + \chi_A N_A$ and $U^M = \left(u_{AB} - \frac{u_{AA} + u_{BB}}{2}\right) N_{AB}^{\#}$ and $S^M = k_B \ln g\left(N_A, N_B, N_{AB}^{\#}\right)$ are the Helmholtz free energy of crystalline solid composed only of N_A atoms A, internal energy of mixing, and entropy of mixing or configurational entropy, respectively. The free energy of mixing is normally written as:

$$F^M = w_{A-B} N_{AB}^{\#} - k_B T \ln g\left(N_A, N_B, N_{AB}^{\#}\right),$$

where $w_{A-B} = u_{AB} - \frac{u_{AA} + u_{BB}}{2}$ is called the interaction parameter between the nearest neighbor atoms A and B. The preferential formation of AA and BB pairs of the nearest neighbors should be, if $w_{A-B} > 0$, and vice versa, and the preferential formation of AB pairs should be, if the condition $w_{A-B} < 0$ is fulfilled.

The chemical potentials of atoms A and B, respectively, are obtained as:

$$\mu_A = \left(\frac{\partial F}{\partial N_A}\right)_{T,N_B}$$

$$= -k_B T \ln \sigma_A + \chi_A + w_{A-B}\frac{\partial N_{AB}^{\#}}{\partial N_A} - k_B T\frac{\partial\left[\ln g\left(N_A, N_B, N_{AB}^{\#}\right)\right]}{\partial N_A}$$

$$= \mu_A^0 + w_{A-B}\frac{\partial N_{AB}^{\#}}{\partial N_A} - k_B T\frac{\partial\left[\ln g\left(N_A, N_B, N_{AB}^{\#}\right)\right]}{\partial N_A},$$

$$\mu_B = \left(\frac{\partial F}{\partial N_B}\right)_{T,N_A}$$

$$= -k_B T \ln \sigma_B + \chi_B + w_{A-B}\frac{\partial N_{AB}^{\#}}{\partial N_B} - k_B T\frac{\partial\left[\ln g\left(N_A, N_B, N_{AB}^{\#}\right)\right]}{\partial N_B}$$

$$= \mu_B^0 + w_{A-B}\frac{\partial N_{AB}^{\#}}{\partial N_B} - k_B T\frac{\partial\left[\ln g\left(N_A, N_B, N_{AB}^{\#}\right)\right]}{\partial N_B},$$

where the Helmholtz free energy is used instead of the Gibbs free energy, and $\mu_A^0 = -k_B T \ln \sigma_A + \chi_A$ is the chemical potential of atoms A in the crystalline solid A. Accordingly, the Helmholtz free energy can be rewritten as:

$$F = F_A + F_B + F^M = \mu_A^0 N_A + \mu_B^0 N_B + U^M - TS^M.$$

If the internal energy does not change after mixing ($U^M = 0$), the regular solution is called an ideal regular solution in which atoms A and B mix randomly. The Helmholtz free energy and chemical potentials of atoms A and B in $A_x B_{1-x}$ ideal regular solution are given, respectively, by:

$$F^I = \mu_A^0 N_A + \mu_B^0 N_B - k_B T\left[\ln\frac{(N_A + N_B)!}{N_A! N_B!}\right]$$

$$= \mu_A^0 N_A + \mu_B^0 N_B + k_B T\left(N_A \ln\frac{N_A}{N_A + N_B} + N_B \ln\frac{N_B}{N_A + N_B}\right),$$

$$\mu_A^I = \mu_A^0 + k_B T \ln\frac{N_A}{N_A + N_B},$$

$$\mu_B^I = \mu_B^0 + k_B T \ln\frac{N_B}{N_A + N_B},$$

where $\frac{(N_A+N_B)!}{N_A! N_B!}$ is the total number of the configurations of atoms. The total number of the configurations is used to describe the entropy of mixing since the maximal item in the sum (3.1.3), representing the partition function of the canonical ensemble, differs insignificantly from the sum of all items in Eqn (3.1.3). The reduced Stirling's formula $\ln N! \approx N \ln N - N$ was utilized to derive the configurational entropy of mixing.

The use of the molar quantities and concentrations to describe the regular solutions is often more convenient than the use of the quantities per atom or per bond and numbers of atoms. The interaction parameter per mole can be obtained from the molar internal energy of mixing of the binary "simple" solution proposed by Guggeheim [2]. The simple solution is also called the strictly regular solution having the random distribution of atoms over the lattice sites [2]. The molar internal energy of mixing of A_xB_{1-x} simple solution is:

$$u^M = \alpha_{A-B}x(1-x),$$

where α_{A-B} is the interaction parameter between A and B atoms per mole. The transition from the interaction parameter per bond to the interaction parameter per mole is:

$$\alpha_{A-B} = z_1 w_{A-B} N_{Av},$$

where N_{Av} is Avogadro's number. The Helmholtz free energy and chemical potentials of atoms A and B per mole of A_xB_{1-x} strictly regular solution are written, respectively, as:

$$f = \mu_A^0 x + \mu_B^0(1-x) + \alpha_{A-B}x(1-x) + RT[x \ln x + (1-x)\ln(1-x)],$$

$$\mu_A = \mu_A^0 + \alpha_{A-B}(1-x)^2 + RT \ln x,$$

$$\mu_B = \mu_B^0 + \alpha_{A-B}x^2 + RT \ln(1-x).$$

Normally, the chemical potential of the i-th component of any solution is:

$$\mu_i = \mu_i^0 + RT \ln a_i = \mu_i^0 + RT \ln \gamma_i x_i$$

where a_i and γ_i are the activity and activity coefficient of the i-th component, respectively. The activity coefficients describe the deviation of the internal energy and entropy of the components from those of the same components in the ideal solution.

The activity coefficients of the components of A_xB_{1-x} binary simple solution are:

$$\gamma_A = \exp\left\{\frac{\alpha_{A-B}(1-x)^2}{RT}\right\},$$

$$\gamma_B = \exp\left\{\frac{\alpha_{A-B}x^2}{RT}\right\}.$$

The ternary and more multicomponent regular solutions may be considered in the same way.

The exact calculation of the number of atomic configurations $g(N, N_A, N_{AB}^{\#})$ is very difficult and, thus, the use of the approximated methods for evaluation of the number of configurations normally is necessary. The model of A_xB_{1-x} binary regular solution is equivalent to the famous Ising model [3]. The Ising model is one of the models used to describe magnetic phenomena in crystalline solids. There are two types of spines in this model as well as two types of atoms in A_xB_{1-x} binary regular solution. The Ising model is a lattice model in which the lattice sites are associated with the spin variables called the spin up and spin down. The internal energy of the magnetic system in the Ising model is:

$$U = -\varepsilon \sum_{ij} \sigma_i \sigma_j - H \sum_{i=1}^{N} \sigma_i,$$

where ε is the interaction energy between the nearest neighbor spins, $\varepsilon > 0$ corresponds to the Ising ferromagnet, $\varepsilon < 0$ corresponds to the Ising anti-ferromagnet, the symbol ij denotes the nearest neighbor pair of the lattice sites, $\sigma_i = \pm 1$ is the spin variable, H is the external magnetic field, and N is the total number of the lattice sites. Accordingly, the results of solving the Ising model can be transformed into the results for A_xB_{1-x} regular solution due to a one-to-one correspondence between the interaction energy between spines ε and interaction parameter w_{A-B} that is given by $\varepsilon \leftrightarrow \frac{w_{A-B}}{2}$. The equivalence of these models is essential, since a lot of results already obtained for the Ising model can be immediately extended to A_xB_{1-x} binary regular solutions. The cluster variation method was chosen from the approximated methods developed for the Ising model to be considered in this book, since the hierarchy of its approximations tends to the exact solution as the cluster size becomes infinite [4].

3.2 MOLECULAR REGULAR SOLUTIONS OF BINARY COMPOUNDS

3.2.1 Ternary Alloys

The crystal structure (zinc blende, wurtzite, or rock salt) of $A_xB_{1-x}C$ and AB_xC_{1-x} ternary semiconductor alloys of binary compounds AC, BC and AB, BC, respectively, consists of two geometrically equivalent sublattices (cationic and anionic) as it is explained in Chapter 1, Section 1.7. One of these sublattices fills with atoms of one type; and another sublattice, called the mixed sublattice, contains atoms of two types. Any lattice site has in the nearest surroundings the lattice sites from another sublattice only. Thus, any atom of the mixed sublattice has in its nearest surroundings atoms of one type. Therefore, an exchange of the lattice sites between atoms from the mixed sublattices does not change the numbers of pairs of the nearest neighbors. The electroneutrality condition demands the equality of the numbers of cations and anions. Thus, the crystal structure of such alloys can be represented as a set of tetrads (for the zinc blende and wurtzite structures) or hexads (for the rock salt structure) of the same type of chemical bonds around atoms belonging to the mixed sublattice. The changes in the arrangement of atoms of the mixed sub-lattice can be represented as a variation in the arrangement of such tetrads or hexads of the same chemical bonds. Therefore, such tetrads or hexads are unchanged objects and can be considered as "molecules" of the constituent binary compounds. Accordingly, these alloys can be represented as alloys of the binary molecules. $A_xB_{1-x}C$ and AB_xC_{1-x} alloys are normally called the "quasibinary" alloys because they contain two types of chemical bonds corresponding to two binary compounds AC, BC and AC, BC, respectively.

The interaction parameters between the compounds in $A_xB_{1-x}C$ and AB_xC_{1-x} ternary alloys in the regular solution model are expressed, respectively, by:

$$w_{AC-BC} = u_{A(C)B} - \frac{u_{A(C)A} + u_{B(C)B}}{2},$$

$$w_{AB-AC} = u_{B(A)C} - \frac{u_{B(A)B} + u_{C(A)C}}{2},$$

where $u_{A(C)B}$ is the interaction energy between atoms A and B with the intermediate atom C. The interaction parameters per mole in $A_xB_{1-x}C$ and AB_xC_{1-x} regular solutions are given, respectively, by:

$$\alpha_{AC-BC} = z_2 w_{AC-BC} N_{Av},$$

$$\alpha_{AB-AC} = z_2 w_{AB-AC} N_{Av}$$

where z_2 is the next nearest coordination number or coordination number of the mixed sublattice; $z_2 = 12$ for the zinc blende structure in which the mixed sublattice is the face centered cubic lattice, wurtzite structure in which the mixed sublattice is the hexagonal close-packed lattice, and rock salt structure in which the mixed sublattice is also the face-centered cubic lattice; and N_{Av} is the Avogadro number. Thus, $A_xB_{1-x}C$ and AB_xC_{1-x} ternary alloys with the zinc blende, wurtzite, and rock salt structures are the same from the regular solution model standpoint.

The Helmholtz free energy of $A_xB_{1-x}C$ regular solution is:

$$F = u_A N_A + u_B N_B + u_C(N_A + N_B) + z_1(u_{AC} N_A + u_{BC} N_B)$$
$$+ \frac{z_2}{2}\left[\left(u_{A(C)A} + u_{C(A)C}\right)N_A + \left(u_{B(C)B} + u_{C(B)C}\right)N_B\right]$$
$$+ w_{AC-BC} N_{AB}^{\#} - Ts_A N_A - Ts_B N_B - Ts_C(N_A + N_B)$$
$$- k_B T \ln g\left(N_A, N_B, N_{AB}^{\#}\right) = F_{AC}^0 + F_{BC}^0 + u^M - Ts^M,$$

where u_A and u_{AC} are the energy of the internal degrees of freedom of atom A and internal energy of bond AC, respectively; s_A is the entropy of the internal degrees of freedom of atom A; $N_{AB}^{\#}$ is the number of AB bonds corresponding to the maximal term of the partition function of the solution;

$$F_{AC}^0 = \left[u_A + u_C + z_1 u_{AC} + \frac{z_2}{2} \left(u_{A(C)A} + u_{C(A)C} \right) - T(s_A + s_C) \right] N_A$$

is the Helmholtz free energy of compound AC; and $U^M = w_{AC-BC} N_{AB}^{\#}$ and $S^M = k_B T \ln g(N_A, N_B, N_{AB}^{\#})$ are the internal energy of mixing and entropy of mixing, respectively.

The chemical potentials of AC and BC molecules in $A_x B_{1-x} C$ regular solution are, correspondingly, obtained as:

$$\mu_{AC} = \left(\frac{\partial F}{\partial N_A} \right)_{T, N_B}$$

$$= -k_B T \ln \sigma_{AC} + \chi_{AC} + w_{AC-BC} \frac{\partial N_{AB}^{\#}}{\partial N_A} - k_B T \frac{\partial \left[\ln g(N_A, N_B, N_{AB}^{\#}) \right]}{\partial N_A}$$

$$= \mu_{AC}^0 + w_{AC-BC} \frac{\partial N_{AB}^{\#}}{\partial N_A} - k_B T \frac{\partial \left[\ln g(N_A, N_B, N_{AB}^{\#}) \right]}{\partial N_A},$$

$$\mu_{BC} = \left(\frac{\partial F}{\partial N_B} \right)_{T, N_A}$$

$$= -k_B T \ln \sigma_{BC} + \chi_{BC} + w_{AC-BC} \frac{\partial N_{AB}^{\#}}{\partial N_B} - k_B T \frac{\partial \left[\ln g(N_A, N_B, N_{AB}^{\#}) \right]}{\partial N_B}$$

$$= \mu_{BC}^0 + w_{AC-BC} \frac{\partial N_{AB}^{\#}}{\partial N_B} - k_B T \frac{\partial \left[\ln g(N_A, N_B, N_{AB}^{\#}) \right]}{\partial N_B},$$

where $\mu_{AC}^0 = -k_B T \ln \sigma_{AC} + \chi_{AC}$ is the chemical potential of the pure compound AC. The $AB_x C_{1-x}$ ternary regular solutions may be obtained in the same way.

3.2.2 Quaternary Alloys

$A_x B_y C_{1-x-y} D$ and $AB_x C_y D_{1-x-y}$ quaternary semiconductor alloys of three binary compounds described in Chapter 1, Section 1.8, as well as $A_x B_{1-x} C$ and $AB_x C_{1-x}$ ternary alloys of two binary compounds, can be considered as regular solutions. In such quaternary alloys the mixed sublattice contains three types of atoms (cations or anions), and another sublattice fills with atoms of one type. $A_x B_y C_{1-x-y} D$ and $AB_x C_y D_{1-x-y}$ alloys are normally called the quasiternary alloys because they contain three types of chemical bonds corresponding to three binary compounds AD, BD, CD and AB, AC, AD, respectively. Any atom of the mixed sublattice has atoms of one type in the nearest surroundings. Hence, an exchange of the lattice sites between atoms belonging to the mixed sublattices does not change the numbers of pairs of the nearest neighbors as well as in $A_xB_{1-x}C$ and AB_xC_{1-x} alloys. The crystal structure of such alloys may be also be presented as a set of tetrads for the zinc blende and wurtzite structures or hexads for the rock salt structure of the same type chemical bonds around atoms belonging to the mixed sublattice. The changes in the arrangement of atoms of the mixed sublattice can be also represented as a variation in the arrangement of such tetrads or hexads of the same chemical bonds. Therefore, such tetrads or hexads are unchanged objects and are considered as "molecules" of the constituent binary compounds. Accordingly, these alloys are described as alloys of the binary molecules.

$A_xB_yC_{1-x-y}D$ and $AB_xC_yD_{1-x-y}$ alloys can be regarded as "quasiternary" regular solutions with the lattice, equivalent to the mixed sublattice of the alloys. Accordingly, all formulas that were deduced for ternary regular solutions of two binary compounds are admissible. The interaction parameters between three binary compounds per "bond" between cations or anions for $A_xB_yC_{1-x-y}D$ and $AB_xC_yD_{1-x-y}$ regular solutions are expressed, respectively, as:

$$w_{AD-BD} = u_{A(D)B} - \frac{u_{A(D)A} + u_{B(D)B}}{2},$$

$$w_{AD-CD} = u_{A(D)C} - \frac{u_{A(D)A} + u_{C(D)C}}{2},$$

$$w_{BD-CD} = u_{B(D)C} - \frac{u_{B(D)B} + u_{C(D)C}}{2},$$

and:

$$w_{AB-AC} = u_{B(A)C} - \frac{u_{B(A)B} + u_{C(A)C}}{2},$$

$$w_{AB-AD} = u_{B(A)D} - \frac{u_{B(A)B} + u_{D(A)D}}{2}$$

$$w_{AC-AD} = u_{C(A)D} - \frac{u_{C(A)C} + u_{D(A)D}}{2}$$

and the interaction parameters per mole are given, respectively, by:

$$\alpha_{AD-BD} = z_2 N_{Av} w_{AD-BD}, \quad \alpha_{AD-CD} = z_2 N_{Av} w_{AD-CD},$$

$$\alpha_{BD-CD} = z_2 N_{Av} w_{BD-CD},$$

and:

$$\alpha_{AB-AC} = z_2 N_{Av} w_{AB-AC}, \quad \alpha_{AB-AD} = z_2 N_{Av} w_{AB-AD},$$

$$\alpha_{AC-AD} = z_2 N_{Av} w_{AC-AD}$$

where z_2 is the next nearest coordination number or the coordination number in the mixed sublattice; $z_2 = 12$ for zinc blende structure, wurtzite structure, and rock salt structure. Accordingly, $A_xB_yC_{1-x-y}D$ and $AB_xC_yD_{1-x-y}$ quaternary alloys as well as $A_xB_{1-x}C$ and AB_xC_{1-x} ternary alloys with the zinc blende, wurtzite, and rock salt structures are the same from the regular solution model standpoint.

References

[1] J.H. Hildebrand, A quantitative treatment of deviations from Raoult's law, Proc. Nat. Acad. Sci. U.S.A. 13 (1927) 267–272.
[2] E.A. Guggenheim, Mixtures, Oxford University Press, Oxford, 1952.
[3] K. Huang, Statistical Mechanics, John Wiley & Sons, Inc., New York, 1963 (Chapter 16).
[4] R. Kikuchi, CVM entropy algebra, Progr. Theoret. Phys. Suppl. 115 (1994) 1–26.

Cluster Variation Method

Exact solutions exist for a very limited number of the lattice models [1]. Moreover, only some of them, and only in few cases, can be applied to consider crystalline semiconductor alloys. Therefore, it is essential to use approximate methods to estimate thermodynamic characteristics and properties of semiconductor alloys. The cluster variation method, developed by Kikuchi [2] in 1951 originally for the Ising ferromagnet with lattices having the first coordination numbers $z_1 > 3$, plays an important role in approximation techniques. Schlijper [3] and Kikuchi [4] showed that the results obtained by using the hierarchies of the approximations tend to the exact solutions for the two- [3] and three-dimensional [4] lattice models. Accordingly, due to the equivalence of the Ising model and the regular solution model [5], this approximate method is one of the most powerful for deriving the regular solutions containing any number of components. The main problems of the cluster variation method are obtaining the approximate expressions for the number of configurations as functions of numbers (concentrations) of clusters (atoms, pairs, triads, etc.) and the following free energy minimizing.

The replacement of the values of the Gibbs free energy by the values of the Helmholtz free energy, and vice versa, established in Chapter 2, Section 2.9 for the considerations of the thermodynamic quantities of condensed matter, is widely used in the current chapter.

4.1 BAKER'S APPROACH

The approach developed by Kikuchi is very difficult to consider large clusters using the approach developed by Kikuchi. Baker [6] elaborated the simple way to describe the configurational entropy for any chosen basic clusters. That is why Baker's approach of the cluster variation method is considered here. The number of configurations is represented as a function of the concentrations of clusters (one-point clusters or single atoms, two-point clusters or atomic pairs of the nearest neighbors, three-point clusters or triads of atoms, etc.). For the sake of brevity, in this section, the general concepts of the cluster variation method are examined for A_xB_{1-x} binary regular solution. The extension of the multicomponent regular solutions such as $A_xB_yC_{1-x-y}$ ternary and $A_xB_yC_zD_{1-x-y-z}$ quaternary can be fulfilled immediately. Each lattice site should be occupied by any atom from a given set of these atomic species A or B atom. The cluster distribution is determined by minimizing the free energy. The largest cluster and other clusters are called, respectively, a basic cluster and subclusters. The larger basic clusters yield to the higher-level approximations that tend to the exact solution. However, the difficulty of minimizing the free energy increases sharply with the enlargement of the basic cluster. Due to this fact, only basic clusters occupying the relatively small numbers of the lattice points are normally treated now. The configurational entropy is calculated as follows. A basic cluster is a group consisting of n atoms in which each atom has at least one nearest neighbor if $n > 1$. A group of m ($m < n$) atoms in which each atom also has at least one nearest neighbor if $m > 1$ is an m-point subcluster. The m-point subcluster having the a-th type of the arrangement of atoms over the lattice sites is called the m_a subcluster. Atom A as well as atom B can be situated over any lattice site. The basic clusters as well as m_a subclusters related by the symmetry operations of the lattice are considered as identical. Any basic cluster configuration as well as any m_a subcluster configuration can be specified by a set of n or m numbers $\{i, j, ..., k\}$, respectively, where $i, j, ..., k$ correspond to A or B atoms or a set of n or m letters A and B. All these configurations $\{i, j, ..., k\}$ of the basic cluster as well as m_a subcluster that are related by symmetry operations of the lattice of this basic cluster or subcluster have the same probability (concentration). The basic cluster and m_a subcluster configurations $k = 1, ..., g$ and $l = 1, ..., h$ generated by the symmetry operations are denoted by $\alpha_k(n)$ and $\alpha_l(m_a)$, respectively.

The molar configurational entropy (entropy of mixing) is written as:

$$s^M = R\delta(n) \sum_k \alpha_k(n)x_k(n)\ln x_k(n) + R\sum_m \delta(m_a) \sum_{m_a} \alpha_l(m_a)x_l(m_a)\ln x_l(m_a),$$

where $\delta(n)$ and $\delta(m_a)$ are the entropy coefficients corresponding to the basic and m_a subcluster, respectively, and $x_k(n)$ and $x_l(m_a)$ are the concentrations of the basic and m_a subcluster in configurations k and l, respectively. The entropy coefficients are obtained as follows:

$$\delta(n) = -\frac{N(n)}{N},$$

$$\delta(m_a) = -\frac{N(m_a)}{N} - \sum_{q_b=m_a+1}^{n} M(m_a, q_b)\delta(q_b), \quad (1 \leq ma \leq n),$$

where N is the total number of atoms, and $N(n)$ and $N(m_a)$ are the numbers of the basic clusters and m_a subclusters, respectively. $M(m_a,q_b)$ is the number of subclusters m_a in cluster q_b ($n \le q > m$). The expressions for the coefficients $\delta(m_a)$ mean that only completely overlapping figures of two adjacent basic clusters can be subclusters. The entropy coefficients of all other m-point ($m < n$) subclusters (not completely overlapping figures of the basic clusters) are equal to zero. It is evident from the following consideration that if subcluster m_c is a partially overlapping figure of two adjacent basic clusters, then all subclusters q_b ($m + 1 \le q \le n - 1$) are also the partially overlapping figures. The entropy coefficients $\delta(n - 1),\dots,\delta(m_c)$ are:

$$\delta(n - 1) = -\frac{N(n - 1)}{N} - M(n - 1, n)\delta(n)$$

$$= -\frac{M(n - 1, n)N(n)}{N} - M(n - 1, n)\delta(n) = 0,$$

$$\delta(n - 2) = -\frac{N(n - 2)}{N} - M(n - 2, n)\delta(n)$$

$$= -\frac{M(n - 2, n)N(n)}{N} - M(n - 2, n)\delta(n) = 0,$$

$$\delta(m_c) = -\frac{N(m_a)}{N} - M(m_c, n)\delta(n) = -\frac{M(m_c, n)N(n)}{N} - M(m_c, n)\delta(n) = 0.$$

Thus, the largest subcluster, q_b, for which the entropy coefficient, $\delta(q_b)$, does not vanish, can be given by the largest completely overlapping figure of two adjacent basic clusters. Next subclusters with the non-vanishing coefficients should be also completely overlapping figures of the basic clusters.

4.2 ONE-POINT APPROXIMATION FOR BINARY REGULAR SOLUTIONS

4.2.1 Helmholtz Free Energy and Chemical Potentials

Atoms are the basic clusters in the one-point approximation equivalent to the strictly regular approximation in the theory of regular solutions. The types, the concentrations, and the numbers of the distinguishable configurations of the clusters are shown in Table 4.1.

The entropy coefficient $\delta(1)$ is written as $\delta(1) = -\frac{N(1)}{N} = -1$. Accordingly, the molar entropy of mixing of A_xB_{1-x} regular solution is

$$s^M = R\delta(1) \sum_{l=1}^{2} \alpha_l(1)x_l(1)\ln x_l(1).$$

TABLE 4.1 One-point clusters

Number	Type	Concentration	Number of distinguishable configurations, $\alpha_i(1)$
1	A	$x_1(1)$	1
2	B	$x_2(1)$	1

The entropy of mixing corresponds to the random arrangement of atoms over the lattice sites. The concentration of AB pairs, when atoms are randomly distributed, and the molar internal energy of mixing are given, respectively, by $x_{AB} = 2x_1(1)x_2(1)$ and $u^M = 2\alpha_{A-B}x_1(1)x_2(1)$, where α_{A-B} is the interaction parameter between A and B atoms. Hence, the molar Helmholtz free energy, the free energy of the components, the free energy of mixing, the chemical potentials, and the activity coefficients of A and B atoms are written, respectively, as:

$$f = f_A^0 + f_B^0 + f^M = \mu_A x + \mu_B(1-x)$$
$$= \mu_A^0 x + \mu_B^0(1-x) + \alpha_{A-B}x(1-x) - RT[x \ln x + (1-x)\ln(1-x)],$$

$$f_A^0 = \mu_A^0 x, \quad f_B^0 = \mu_B^0(1-x),$$

$$f^M = u^M - Ts^M = \alpha_{A-B}x(1-x) - RT[x \ln x + (1-x)\ln(1-x)],$$

$$\mu_A = \mu_A^0 + RT \ln \gamma_A x,$$

$$\mu_B = \mu_B^0 + RT \ln \gamma_B(1-x),$$

$$\gamma_A = \exp\left\{\frac{\alpha_{A-B}(1-x)^2}{RT}\right\} \quad \text{and} \quad \gamma_B = \exp\left\{\frac{\alpha_{A-B}x^2}{RT}\right\},$$

where μ_A^0 is the molar chemical potential of the substance A. If composition $x \ll 1$, a regular solution is called dilute or ultra dilute depending on a value of x. The molar chemical potentials of A and B atoms in such solutions are:

$$\gamma_A = \exp\left\{\frac{\alpha_{A-B}(1-x)^2}{RT}\right\} \approx \exp\left\{\frac{\alpha_{A-B}}{RT}\right\},$$

$$\gamma_B = \exp\left\{\frac{\alpha_{A-B}x^2}{RT}\right\} \approx 1.$$

4.2.2 Miscibility Gap

Immiscibility is a serious problem in semiconductor alloys growing near thermodynamic equilibrium. The internal energies of mixing of all semiconductor alloys are positive. The tendency to phase separation of an alloy arises from this positive energy. Accordingly, any homogeneous semiconductor alloy forms near thermodynamic equilibrium due to the occurrence of configurational entropy. The Helmholtz free energy of A_xB_{1-x} regular solution is written as $f = f_A + f_B + u^M - Ts^M$, where f_A, u^M, and s^M are the free energy of component A, the internal energy of mixing, and the entropy of mixing, respectively. The internal energy of mixing of a regular solution is independent of temperature, but the configurational entropy term is directly proportional to the absolute temperature. Therefore, a decrease of temperature increases the relative contribution of the internal energy of mixing to the free energy. This can lead to the phase separation of the homogeneous alloy into a heterogeneous system consisting of two regular solutions (decomposed regular solution). The temperature at which the free energies of the homogeneous and decomposed regular solutions become equal is called a temperature of the miscibility gap boundary. At the lower temperatures, the free energy of the homogeneous regular solution is larger than the free energy of the decomposed solution. The transformation of a homogeneous solution into a decomposed solution changes only the internal energy of mixing and entropy of mixing, and only these terms have to be considered when calculating the miscibility gap boundary. The free energy of mixing of the homogeneous regular solution given by:

$$f^M = u^M - Ts^M = \alpha_{A-B}x(1 - x) + RT[x \ln x + (1 - x)\ln(1 - x)].$$

is the symmetrical function relatively to composition $x = \frac{1}{2}$. Accordingly, the free energy of mixing of A_xB_{1-x} solution after its decomposition is:

$$f^{MD} = \gamma f^M(x_1) + (1 - \gamma)f^M(x_2) = \gamma f^M(1 - x_1) + (1 - \gamma)f^M(1 - x_2),$$

(4.2.2.1)

where x_1 and x_2 are, respectively, the concentrations of atoms A in the first and second phases and γ is the fraction of the first phase in the decomposed solution. It follows from Eqn (4.2.2.1) that $x_2 = 1 - x_1$ and $f^{MD} = f^M(x_1)$. Thus, the free energy of mixing of the decomposed solution is equal to the free energy of mixing of the homogeneous solution with the composition corresponding to the miscibility gap boundary.

The miscibility gap boundary is obtained by minimizing the free energy of mixing of the decomposed solution:

$$\frac{df^M}{dx} = \alpha_{A-B}(1 - 2x) + RT\ln\frac{x}{1 - x} = 0. \qquad (4.2.2.2)$$

4.2.3 Spinodal Decomposition Range

Spinodal decomposition occurs in a regular solution when it losses stability with respect to the negligibly small phase separation perturbations (composition fluctuations). As stated by Gibbs [7], spinodal decomposition begins from composition changes that are small in degree but large in extent. The boundary of the spinodal decomposition range of A_xB_{1-x} regular solution is obtained as follows. The equal fractions of A and B atoms δN_A and δN_B over the nearest neighbor lattice sites participate in the exchange of the lattice sites. Thus, as a result of such exchange the decomposed solution, consisting of two domains with the different compositions and situated into the homogeneous solution, forms. The volumes of the occurred domains of the decomposed solution should be equal to each other, since atoms situated over the nearest neighbor lattice sites participate in such exchange. The variation of the free energy of mixing during the initial stage of spinodal decomposition is:

$$\delta f^M = \frac{1}{2}f^M(x + \delta x) + \frac{1}{2}f^M(x - \delta x) - f^M(x),$$

where f^M are the free energies of mixing of the homogeneous solution, and $x + \delta x$ and $x - \delta x$ are the compositions of the first and second formed domains. The free energy of mixing of the homogeneous solution is written as:

$$f^M(x) = \alpha_{A-B}x(1 - x) + RT[x\ln x + (1 - x)\ln(1 - x)],$$

The variation of the free energy of mixing obtained by using the Taylor's series expansion is:

$$\delta f^M = \frac{1}{2}\frac{d^2f^M(x)}{dx^2}(\delta x)^2$$

The regular solution losses stability if the condition:

$$\delta f^M = 0$$

is fulfilled, since, if $\delta f^M < 0$, then the negligibly small phase separation perturbation decreases the free energy. Accordingly, the boundary of the spinodal decomposition range of a regular solution is obtained from the equation:

$$\frac{d^2 f^M}{dx^2} = -2\alpha_{A-B} + \frac{RT}{x(1-x)} = 0. \tag{4.2.3.1}$$

The lattice parameters of components of semiconductor alloys are normally different, whereas the lattice parameters of components of the regular solutions are supposed to be the same. This difference causes the coherency strain energy in the decomposed semiconductor alloy containing lattice-mismatched components, since the formed domains are in the strained state. The coherency strain energy increasing the internal energy hinders spinodal decomposition. The concept of the coherency strain energy significantly improving the description of the spinodal decomposition ranges was introduced by Cahn for isotropic solid solutions and alloys with cubic structures [8,9]. Later, Stringfellow [10] developed this approach to derive the spinodal decomposition ranges of the III–V semiconductor alloys with the zinc blende structure.

For many semiconductor alloys of the lattice-mismatched constituents, spinodal decomposition is unprofitable at any temperature since it leads to an increase of the free energy due to the coherency strain energy. The spinodal decomposition range obtained with and without taking into account the coherency strain energy (curves "SD range with CSE" and "SD range without CSE", respectively) and miscibility gap (curve "MG") of $A_x B_{1-x}$ semiconductor alloy of the lattice-mismatched constituents are shown in Figure 4.1.

The spinodal decomposition range obtained without taking into account the coherency strain energy (Eqn (4.2.2.2)):

$$T = \frac{2\alpha_{A-B} x(1-x)}{R}.$$

and miscibility gap (Eqn (4.2.2.1)):

$$T = \frac{\alpha_{A-B}(2x-1)}{R \ln \frac{x}{1-x}}$$

are the symmetric functions relatively to alloy composition $x = \frac{1}{2}$. The temperatures of this spinodal decomposition boundary and miscibility gap are equal for this composition. The spinodal decomposition range derived considering the coherency strain energy (Eqn (4.2.3.2)):

$$T = \frac{2x(1-x)}{R\left[\alpha_{A-B} - \frac{(C_{11}-C_{12})(C_{11}+2C_{12})}{C_{11}}\left(\frac{a_A-a_B}{a}\right)^2 v\right]}$$

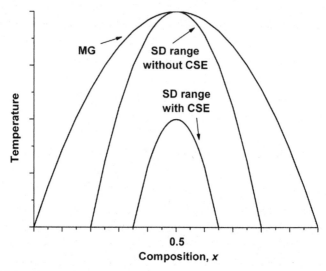

FIGURE 4.1   The miscibility gap and spinodal decomposition ranges estimated with and without the coherency strain energy.

in the general case, is the asymmetric function relatively to alloy composition $x = \frac{1}{2}$. The states of the alloys above the curve "MG" are stable with respect to decomposition, since in these states the Helmholtz free energy attains its absolute minimum. The states of the alloys lying between the curves "MG" and "SD with CSE" are metastable with respect to decomposition as the Helmholtz free energy reaches its relative minimum. Accordingly, the finite decomposition fluctuation may decrease the free energy. Such fluctuation should develop. The states of the alloys under the curve "SD with CSE" are unstable with respect to decomposition since a negligibly small decomposition fluctuation decreases the free energy and has to develop.

The initial stage of spinodal decomposition is accompanied by self-diffusion transfers of atoms on the distances equal to the distances between the nearest neighbor atoms. In crystalline semiconductor alloys, spinodal decomposition should occur in the planes ensuring the minimal coherency strain energy between the formed domains. Thus, the initial stage of spinodal decomposition has to be presented as a formation of two very thin layers. The compositions of the layers at the initial stage can be considered as constant values due to their small thickness. As the spinodal decomposition is developed, the transfers of atoms and thicknesses of the layers become larger, and the composition of the layers varies with a distance. Afterward, the difference in the average concentrations of the components in the domains increases continuously.

The coherency strain energy at the initial stage of spinodal decomposition is obtained as follows. The molar elastic energy of the lattice mismatched thin layer in a crystalline solid is:

$$u = \frac{1}{2} v \left(2\sigma_{\parallel} \varepsilon_{\parallel} + \sigma_{\perp} \varepsilon_{\perp} \right),$$

where v; $\sigma_{\parallel}$, $\sigma_{\perp}$; and $\varepsilon_{\parallel}$, $\varepsilon_{\perp}$ are the molar volume and the parallel and perpendicular to the layer–substrate interface stresses and strains, respectively. Accordingly, the elastic energy of the decomposed alloy or the coherency strain energy can be written similar to the elastic energy of two lattice-mismatched layers that is given by:

$$u^C = \frac{1}{4} \sum_{i=1}^{2} v_i \left(2\sigma_{i\parallel} \varepsilon_{i\parallel} + \sigma_{i\perp} \varepsilon_{i\perp} \right),$$

where v_i; $\sigma_{i\parallel}$, $\sigma_{i\perp}$; and $\varepsilon_{i\parallel}$, $\varepsilon_{i\perp}$ are the molar volume and the parallel and perpendicular to the layer–crystal interface stresses and strains of the i-th ($i = 1, 2$) domain of the decomposed alloy, respectively. The formation of the decomposed layers in the crystalline alloys should be in the planes where their strain energy is minimal. As shown in Appendix 1, in crystalline solids with the cubic structure, the minimal strain energy occurs in the lattice-mismatched layers oriented in the {001} planes. Thus, the initial stage of spinodal decomposition in semiconductor alloys should be the formation of two thin layers oriented in one of the {001} planes. The axis perpendicular to the layer–homogeneous alloy interface is characterized by vanishing forces in this direction, $\sigma_{\perp} = 0$. Therefore, the parallel to the layer–homogeneous alloy interface stresses can be written as:

$$\sigma_{\parallel} = (C_{11} - C_{12})(C_{11} + 2C_{12})/C_{11},$$

where C_{11} and C_{12} are the stiffness coefficients, which are described by the linear functions of the concentrations of the components. The coherency strain energy of the two layers can be rewritten as:

$$u^C = \frac{1}{2} \sum_{i=1}^{2} \frac{\left(C_{11}^i - C_{12}^i \right) \left(C_{11}^i + 2C_{12}^i \right)}{C_{11}^i} \left(\frac{a_i - a}{a} \right)^2 v_i,$$

where C_{11}^i, a_i, a, and v_i are the stiffness coefficient, the lattice parameter of the i-th layer, the lattice parameter of the homogeneous alloy, and the molar volume of the i-th layer, respectively. The molar volume, the stiffness coefficient, and lattice parameter of the i-th layer are almost equal, respectively, to the molar volume, the stiffness coefficient, and lattice parameter of the homogeneous alloy. Therefore, in the last formula of the coherency strain energy all multipliers except the last weakly depend on the composition. Accordingly, in the calculations, the molar volumes and stiffness coefficients of the layers can be considered as the same quantities of the homogeneous alloy, and the coherency strain energy can be rewritten as:

$$u^C = \frac{(C_{11} - C_{12})(C_{11} + 2C_{12})}{C_{11}} \left[\frac{a(x + \delta x) - a(x)}{a(x)} \right]^2 v.$$

The strains in the homogeneous alloy were neglected when deducing the expression of the coherency strain energy since the dimensions of the homogeneous alloy are much larger than the thickness of the layers. The free energy of mixing of the decomposed alloy is:

$$f^M = \frac{1}{2} f^M(x + \delta x) + \frac{1}{2} f^M(x - \delta x) + u^C.$$

The variation of the free energy of mixing after the formation of the layers is given by:

$$\delta f^M = \frac{1}{2} f^M(x + \delta x) + \frac{1}{2} f^M(x - \delta x) - f^M(x) + u^C.$$

The alloy loses stability if the condition $\delta f^M = 0$ is fulfilled. The coherency strain energy obtained by using the Taylor's series expansion is:

$$u^C = \frac{1}{2} \frac{d^2 u^C}{dx^2} (\delta x)^2.$$

Finally, the spinodal decomposition range of $A_x B_{1-x}$ binary semi-conductor alloy of the lattice-mismatched components is given by:

$$-\alpha_{A-B} + \frac{RT}{2x(1-x)} + \frac{(C_{11} - C_{12})(C_{11} + 2C_{12})}{C_{11}} \left(\frac{a_A - a_B}{a} \right)^2 v = 0, \quad (4.2.3.2)$$

where $C_{11} = C_{11}^A x + C_{11}^B (1 - x)$, $a = a_A x + a_B (1 - x)$ and $v = v_A x + v_B (1 - x)$ are the stiffness coefficient and the molar volume of the homogeneous alloy, respectively; and C_{11}^A, v_A, and a_A, accordingly, are the stiffness coefficient, the molar volume, and the lattice parameter of the semiconductor A.

The orientation of the occurred lattice-mismatched layers, causing the minimal strain energy in alloys with the wurtzite structure, depends on the stiffness coefficients and lattice parameters of the components and also on the alloy composition, as shown in Appendix 2.

4.3 ONE-POINT APPROXIMATION FOR TERNARY REGULAR SOLUTIONS

4.3.1 Helmholtz Free Energy and Chemical Potentials

The types, the concentrations, and the numbers of the different configu-rations of the clusters in $A_x B_y C_{1-x-y}$ regular solution are shown in Table 4.2.

TABLE 4.2 One-point clusters

Number	Type	Concentration	Number of different configurations, $\alpha_l(1)$
1	A	$x_1(1)$	1
2	B	$x_2(1)$	1
3	C	$x_3(1)$	1

The entropy coefficient $\delta(1)$ is given by:

$$\delta(1) = -\frac{N(1)}{N} = -1.$$

Consequently, the molar entropy of mixing of $A_x B_y C_{1-x-y}$ regular solution is:

$$s^M = R\delta(1)\sum_{l=1}^{3}\alpha_l(1)x_l(1) - R[x \ln x + y \ln y + (1 - x - y)\ln(1 - x - y)],$$

$$(4.3.3.1)$$

where $x = N_A/N$, $y = N_B/N$, and $N = N_A + N_B + N_C$. The configurational entropy (Eqn (4.3.3.1)) corresponds to the random arrangement of atoms on the lattice sites. Such distribution of atoms is described by the following formulas for the concentrations of pairs AB, AC, BC and internal energy of mixing, that are given, respectively, by:

$$x_{AB} = xy, \quad x_{AC} = x(1 - x - y), \quad x_{BC} = y(1 - x - y)$$

$$u^M = \alpha_{A-B}xy + \alpha_{A-C}x(1 - x - y) + \alpha_{B-C}y(1 - x - y).$$

The free energy of mixing of the ternary regular solution $f^M = u^M - Ts^M$ in contrast to the free energy of mixing of $A_x B_{1-x}$ binary regular solution is the asymmetric function of the composition. The Helmholtz free energy is:

$$f = \mu_A^0 x + \mu_B^0 y + \mu_C^0(1 - x - y) + f^M,$$

where μ_A^0 is the chemical potential of the substance A. The chemical potentials of A, B, and C atoms, obtained in the same manner as the chemical potentials of the components of $A_x B_{1-x}$ binary regular solution, are given, respectively, by:

$$\mu_A = \mu_A^0 + RT \ln x + \alpha_{A-B}(1 - x)y + \alpha_{A-C}(1 - x)(1 - x - y)$$
$$- \alpha_{B-C}y(1 - x - y),$$

$$\mu_B = \mu_B^0 + RT \ln y + \alpha_{A-B}x(1 - y) - \alpha_{A-C}x(1 - x - y)$$
$$+ \alpha_{B-C}(1 - y)(1 - x - y),$$

$$\mu_C = \mu_C^0 + RT \ln(1 - x - y) - \alpha_{A-B}xy + \alpha_{A-C}x(x + y) + \alpha_{B-C}y(x + y).$$

The activity coefficients of A, B, and C atoms in $A_xB_yC_{1-x-y}$ regular solution are written, respectively, as:

$$\gamma_A = \exp\left\{\frac{\alpha_{A-B}(1 - x)y + \alpha_{A-C}(1 - x)(1 - x - y) - \alpha_{B-C}y(1 - x - y)}{RT}\right\},$$

$$\gamma_B = \exp\left\{\frac{\alpha_{A-B}x(1 - y) - \alpha_{A-C}x(1 - x - y) + \alpha_{B-C}(1 - y)(1 - x - y)}{RT}\right\},$$

$$\gamma_C = \exp\left\{\frac{-\alpha_{A-B}xy + \alpha_{A-C}x(x + y) + \alpha_{B-C}y(x + y)}{RT}\right\}.$$

If the composition of the regular solution corresponds to the condition $x_i \to 0$, $(i = A, B)$, $x_C \to 1$, then it is called the dilute or ultra-dilute solution, and activity coefficients of its components are:

$$\gamma_A \approx \exp\left\{\frac{\alpha_{A-C}}{RT}\right\}, \quad \gamma_B \approx \exp\left\{\frac{\alpha_{B-C}}{RT}\right\}, \quad \gamma_C \approx 1.$$

4.3.2 Miscibility Gap

The miscibility gap is derived by minimizing the Helmholtz free energy of the two-phase system represented as a canonical ensemble. The Helmholtz free energy of mixing changes only as a result of phase separation of $A_xB_yC_{1-x-y}$ homogeneous regular solution. Therefore, this part of the free energy is taken into account. The free energy of mixing of the two-phase system is:

$$F^M = \sum_{i=1}^{2} F_i^M(N, N_A, N_B, N_i, N_{iA}, N_{iB}),$$

where $N = N_A + N_B + N_C$, N_A, and N_B are given values and N_i, N_{iA}, and N_{iB} are the numbers of atoms in i-th phases that are reciprocally dependent variables. Therefore, there are three constraints that are written as:

$$\varphi_1 = N_1 + N_2 - N = 0, \quad \varphi_2 = N_{1A} + N_{2A} - N_A = 0,$$

$$\varphi_3 = N_{1B} + N_{2B} - N_B = 0.$$

The minimum condition obtained by the method of the Lagrange undetermined multipliers (Appendix 3) is done by a system of equations:

$$\frac{\partial L}{\partial N_i} = 0, \quad \frac{\partial L}{\partial N_{iA}} = 0, \quad \frac{\partial L}{\partial N_{iB}} = 0, \quad \varphi_j = 0, \quad (i = 1, 2; j = 1, ..., 3),$$

where $L = F^M + \sum_{j=1}^{3} \lambda_j \varphi_j$, λ_j are the Lagrange undetermined multipliers. The free energy of mixing of the i-th phase is expressed by:

$$F_i^M = w_{A-B} \frac{z_1 N_{iA} N_{iB}}{N_i} + w_{A-C} \frac{z_1 N_{iA}(N_i - N_{iA} - N_{iB})}{N_i}$$

$$+ w_{B-C} \frac{z_1 N_{iB}(N_i - N_{iA} - N_{iB})}{N_i}$$

$$+ k_B T \left[N_{iA} \ln \frac{N_{iA}}{N_i} + N_{iB} \ln \frac{N_{iB}}{N_i} \right.$$

$$\left. + (N_i - N_{iA} - N_{iB}) \ln \frac{N_i - N_{iA} - N_{iB}}{N_i} \right].$$

Finally, the system of the equations deriving the miscibility gap boundary is found to be:

$$\sum_{i=1}^{2} (-1)^i [-\alpha_{A-B} x_i y_i + \alpha_{A-C} x_i (x_i + y_i) + \alpha_{B-C} y_i (x_i + y_i)$$

$$+ RT \ln(1 - x_i - y_i)] = 0,$$

$$\sum_{i=1}^{2} (-1)^i \left[\alpha_{A-B} y_i + \alpha_{A-C}(1 - 2x_i - y_i) - \alpha_{B-C} y_i + RT \ln \frac{x_i}{1 - x_i - y_i} \right] = 0,$$

$$\sum_{i=1}^{2} (-1)^i \left[\alpha_{A-B} x_i - \alpha_{A-C} x_i + \alpha_{B-C}(1 - x_i - 2y_i) + RT \ln \frac{y_i}{1 - x_i - y_i} \right] = 0,$$

where $x_2 = \frac{x - \gamma x_1}{1 - \gamma}$, $y_2 = \frac{y - \gamma y_1}{1 - \gamma}$, $\gamma = \frac{N_1}{N}$, and x_1, y_1, γ are the independent variables.

4.3.3 Spinodal Decomposition Range

The spinodal decomposition range of $A_x B_y C_{1-x-y}$ ternary regular solution is derived by the equality condition between the free energies of the phase-separated solution with the negligibly small composition perturbation and the homogeneous solution. An exchange of atoms between domains 1 and 2 in the perturbation can be represented as:

$$A(1) \leftrightarrow B(2), \quad A(1) \leftrightarrow C(2)$$

or as a simultaneous exchange of atoms A from region 1 with B and C atoms of region 2. The variation of the Helmholtz free energy after the initial stage of spinodal decomposition is:

$$\delta f = \frac{1}{2} f^M(x - \delta x_1 - \delta x_2, y + \delta x_1) + \frac{1}{2} f^M(x + \delta x_1 + \delta x_2, y - \delta x_1) - f^M(x, y),$$

where $\delta_1 x$ and $\delta_2 x$ are the concentration fractions of B and C atoms participating in the exchange. The variation of the free energy obtained by using the Taylor's multidimensional series expansion has the quadratic form:

$$\delta f^M = \frac{1}{2}\frac{\partial^2 f^M}{\partial x^2}(\delta x_1 + \delta x_2)^2 - \frac{\partial^2 f^M}{\partial x \partial y}(\delta x_1 + \delta x_2)\delta x_1 + \frac{1}{2}\frac{\partial^2 f^M}{\partial y^2}(\delta x_1)^2.$$

The spinodal decomposition range of $A_x B_y C_{1-x-y}$ ternary regular solution is derived by the condition:

$$\delta f^M = 0.$$

In accordance with Sylvester's criterion [11], the quadratic form ceases to be a positive definite when one of the two quantities:

$$\frac{\partial^2 f^M}{\partial x^2},$$

$$\begin{vmatrix} \dfrac{\partial^2 f^M}{\partial x^2} & -\dfrac{\partial^2 f^M}{\partial x \partial y} \\[2ex] -\dfrac{\partial^2 f^M}{\partial x \partial y} & \dfrac{\partial^2 f^M}{\partial y^2} \end{vmatrix}$$

becomes equal to zero.

As explained in Section 4.2.3, spinodal decomposition leads to the occurrence of the coherency strain energy in semiconductor alloys with lattice-mismatched components. Thus, the new domains formed are in the strained state. The initial stage of spinodal decomposition results in the self-diffusion transfers of atoms on the distances of the order of a lattice parameter. In crystalline semiconductor alloys, the initial stage of spinodal decomposition results in the transfers of atoms in the planes ensuring the minimal coherency strain energy. Subsequently, two very thin layers form in the homogeneous alloy. The compositions of the formed layers at the initial stage are considered as constant values due to their small thickness. In ternary semiconductor alloys with the cubic structure, the occurred coherency strain energy is given by:

$$u^C = \frac{(C_{11} - C_{12})(C_{11} + 2C_{12})}{C_{11}}\left(\frac{a(x - \delta x_1 - \delta x_2, y + \delta x_1) - a}{a}\right)^2 v,$$

where $C_{ij} = C_{ij}^A x + C_{ij}^B y + C_{ij}^C(1 - x - y)$, v, a, and C_{ij}^A are the stiffness coefficients, the molar volume, and the lattice parameter of the homogeneous alloy, respectively, and the stiffness coefficient of semi-conductor A. The free energy of mixing of the decomposed alloy is:

$$f^M = \frac{1}{2}f^M(x - \delta x_1 - \delta x_2, y + \delta x_1) + \frac{1}{2}f^M[x + (\delta x_1 + \delta x_2), y - \delta x_1] + u^C,$$

The variation of the free energy of mixing is:

$$\delta f^M = \frac{1}{2}f^M(x - \delta x_1 - \delta x_2, y + \delta x_1)$$
$$+ \frac{1}{2}f^M[x + (\delta x_1 + \delta x_2), y - \delta x_1] - f^M(x, y) + u^C.$$

The alloy losses stability if the condition:

$$\delta f^M = 0$$

is fulfilled. The coherency strain energy obtained by using the Taylor's series expansion is given as:

$$u^C = \frac{1}{2}\frac{\partial^2 u^C}{\partial x^2}(\delta x_1 + \delta x_2)^2 - \frac{\partial^2 u^C}{\partial x \partial y}(\delta x_1 + \delta x_2)\delta x_1 + \frac{1}{2}\frac{\partial^2 u^C}{\partial y^2}(\delta x_1)^2.$$

The spinodal decomposition range of a semiconductor alloy is derived by the condition $\delta f^M = 0$, which is fulfilled when one of the following two expressions becomes equal to zero:

$$-2\alpha_{A-C} + \frac{RT(1-y)}{x(1-x-y)} + 2\frac{(C_{11} - C_{12})(C_{11} + 2C_{12})}{C_{11}}\left(\frac{a_A - a_C}{a}\right)^2 v,$$

$$\left[-2\alpha_{A-C} + \frac{RT(1-y)}{x(1-x-y)} + 2\frac{(C_{11} - C_{12})(C_{11} + 2C_{12})}{C_{11}}\left(\frac{a_A - a_C}{a}\right)^2 v\right]$$

$$\times \left[-2\alpha_{B-C} + \frac{RT(1-x)}{y(1-x-y)} + 2\frac{(C_{11} - C_{12})(C_{11} + 2C_{12})}{C_{11}}\left(\frac{a_B - a_C}{a}\right)^2 v\right]$$

$$-\left[\alpha_{A-B} - \alpha_{A-C} - \alpha_{B-C} + \frac{RT}{1-x-y}\right.$$

$$\left.+ 2\frac{(C_{11} - C_{12})(C_{11} + 2C_{12})}{C_{11}}\frac{(a_A - a_C)(a_B - a_C)}{a^2}v\right]^2,$$

where $v = v_A x + v_B y + v_C(1 - x - y)$, $C_{11} = C_{11}^A x + C_{11}^B y + C_{11}^C(1 - x - y)$, and $a = a_A x + a_B y + a_C(1 - x - y)$, are the molar volume, the stiffness coefficient, and the lattice parameter of the homogeneous alloy, respectively. The consideration of the other processes of the simultaneous exchange of atoms leads to the same result. The extension of such considerations of spinodal decomposition to more multicomponent systems is straightforward.

4.4 TWO-POINT APPROXIMATION FOR BINARY REGULAR SOLUTIONS

4.4.1 Helmholtz Free Energy and Short-Range Order

The two-point approximation of the cluster variation method is equivalent to the quasi-chemical approximation in the theory of regular solutions. The basic clusters in the two-point approximation are the pairs of the nearest neighbor atoms (pairs of the nearest neighbor lattice points), since each lattice point is occupied by an atom. There are three types of pairs (AA, BB, and AB) in A_xB_{1-x} regular solution. The cluster variation method is based on the supposition that the basic clusters of atoms are capable of freely and randomly mixing. In some cases, the pairs of atoms are also called the bonds between the nearest neighbors. Thus, the pairs of atoms (bonds) are considered as the configurationally independent objects. The entropy of mixing is derived from this supposition. The pair AB can be disposed on the same lattice sites by two configurationally different variants as the pair AB and pair BA, or the number of different configurations that can be generated by the symmetry operations of the basic cluster AB is equal to two. Therefore, one-half of AB pairs are considered as clusters AB and the other half as clusters BA. An atom or a lattice point as an overlapping figure of two basic clusters is a subcluster in the two-point approximation.

The types, the concentrations, and numbers of the different configurations of the basic clusters and subclusters are shown in Tables 4.3 and 4.4.

TABLE 4.3 Two-point clusters

Number	Type	Concentration	Number of different configurations, $\alpha_i(2)$
1	AA	$x_1(2)$	1
2	AB	$x_2(2)$	2
3	BB	$x_3(2)$	1

TABLE 4.4 One-point clusters

Number	Type	Concentration	Number of different configurations, $\alpha_i(1)$
1	A	$x_1(1)$	1
2	B	$x_2(1)$	1

The concentrations of the basic clusters are considered as variables. The concentrations of the subclusters are functions of the concentrations of the basic clusters:

$$x_1(1) = x_1(2) + x_2(2) \quad \text{and} \quad x_2(1) = x_2(2) + x_3(2).$$

There is the constraint between the variables expressed as:

$$x_1(2) + 2x_2(2) + x_3(2) = 1.$$

If the composition of $A_x B_{1-x}$ regular solution is considered as a given value, there is the additional constraint between the variables:

$$x_1(2) + x_2(2) = x.$$

The Helmholtz free energy of mixing is given as $f^M = u^M - Ts^M$, where the internal energy of mixing is:

$$u^M = \alpha_{A-B} x_2(2) = \frac{1}{2} \alpha_{A-B} x_{AB}.$$

The entropy coefficients for the basic clusters and the subclusters in the expression of the entropy of mixing are given, respectively, as:

$$\delta(2) = -\frac{N(2)}{N} = -\frac{z_1}{2},$$

$$\delta(1) = -\frac{N(1)}{N} - M(1,2)\delta(2) = z_1 - 1.$$

Accordingly, the entropy of mixing per mole will have the form:

$$s^M = R\left\{ -\frac{z_1}{2}[x_1(2)\ln x_1(2) + 2x_2(2)\ln x_2(2) + x_3(2)\ln x_3(2)] \right.$$
$$\left. + (z_1 - 1)[x_1(1)\ln x_1(1) + x_2(1)\ln x_2(1)] \right\}.$$

Thus, the free energy of mixing is the function of one independent variable, $x_2(2)$, if the composition is a given value, since $x_1(2) = x_1(1) - x_2(2)$, $x_3(2) = 1 - x_1(1) - x_2(2)$, $x_2(1) = 1 - x_1(1)$, and $x_1(1) \equiv x$.

The concentrations of AB pairs or the short-range order at a given composition is determined by minimizing the free energy of mixing:

$$\frac{df^M}{dx_2(2)} = 0 \quad \text{or}$$

$$\exp\left\{ \frac{2\alpha_{A-B}}{z_1 RT} \right\} = \frac{[x - x_2(2)][1 - x - x_2(2)]}{[x_2(2)]^2}.$$

The concentration of AB pairs is given by:

$$x_{AB} = 2x_2(2) = \frac{\sqrt{1 + 4(\eta^2 - 1)x(1 - x)} - 1}{\eta^2 - 1},$$

where $\eta = \exp\left\{\frac{\alpha_{A-B}}{z_1 RT}\right\}$. Thus, the concentration of AB pairs in the two-point approximation is not only the function of the composition but the temperature as well.

In the theory of regular solutions, the entropy of mixing in the quasi-chemical approximation is obtained as follows. The number of geometrically different configurations $g(N_A, N_B N_{AB}^{\#})$ by using the supposition on the random arrangement of pairs of the nearest neighbors is represented as:

$$g(N_A, N_B, N_{AB}^{\#}) = \varphi(N_A, N_B, N_{AB}^{\#})h(N_A, N_B),$$

where $\varphi(N_A, N_B, N_{AB}^{\#})$ is the number of all configurations of AA, AB, and BB pairs considered as independent objects that can be freely mixed over the lattice sites, and $h(N_A, N_B)$ is the normalization factor. The number of the configurations of the freely mixing pairs is:

$$\varphi(N_A, N_B, N_{AB}^{\#}) = \frac{\left(z\frac{N_A + N_B}{2}\right)!}{\left(\frac{zN_A - N_{AB}^{\#}}{2}\right)!\left(\frac{N_{AB}^{\#}}{2}\right)!\left(\frac{N_{AB}^{\#}}{2}\right)!\left(\frac{zN_B - N_{AB}^{\#}}{2}\right)!}.$$

Two factors $\left(\frac{N_{AB}^{\#}}{2}\right)!$ are used, since the AB and BA pairs are configurationally different. The number of configurations of the pairs over the lattice sites $\varphi(N_A, N_B, N_{AB}^{\#})$ is much more than the number of configurations of atoms in the lattice. This follows from the assumption that the pairs are independent objects. The number of such pairs is equal to the number of bonds between the nearest neighbor atoms. Accordingly, the ratio between the number of pairs and number of atoms is written as $\frac{z_1}{2}$, where z_1 is the coordination number of the nearest neighbors. The total number of configurations of atoms over the lattice sites is:

$$\frac{(N_A + N_B)!}{N_A! N_B!}.$$

Therefore, the normalization factor $h(N_A, N_B)$ not depending on the quantities of pairs is introduced in the expression for the number of configurations. This factor is written as:

$$h(N_A, N_B) = \frac{(N_A + N_B)!}{N_A! N_B!} \times \frac{\left[\frac{z_1 N_A^2}{2(N_A + N_B)}\right]!\left\{\left[\frac{z_1 N_A N_B}{2(N_A + N_B)}\right]!\right\}^2\left[\frac{z_1 N_B^2}{2(N_A + N_B)}\right]!}{\left(z_1 \frac{N_A + N_B}{2}\right)!}.$$

The factor $h(N_A,N_B)$ equalizes the number of configurations of the pairs at the infinitely large temperature with the total number of arrangements of atoms, since the atoms are distributed randomly at the infinitely large temperature. Thus, the normalization factor provides the correct value of the entropy of mixing at the infinitely large temperature. For finite temperatures, the normalization factor leads to underestimated values of the entropy of mixing.

The free energy of mixing of $A_x B_{1-x}$ regular solution in the quasi-chemical approximation is:

$$F^M = U^M - TS^M = w_{A-B}N_{AB} - k_B T \ln\left[h(N_A, N_B)\varphi\left(N_A, N_B, N_{AB}^{\#}\right)\right].$$

The number of AB pairs is calculated by minimizing the free energy of mixing, given by:

$$\frac{dF^M}{dN_{AB}} = 0, \quad \text{or} \quad w_{A-B} - k_B T \frac{d}{dN_{AB}} \ln \varphi\left(N_A, N_B, N_{AB}^{\#}\right).$$

Thus, the number of AB pairs does not depend on the normalization factor. The minimum free energy:

$$\frac{dF^M}{dN_{AB}} = 0$$

corresponds to one allowed reaction between pairs at the rearrangement of atoms over the lattice sites given by $AA + BB = 2AB$. In fact, this reaction is a reaction between bonds similar to the chemical reaction between molecules. That is why this approximation was called quasi-chemical. All other reactions between the pairs (bonds), such as:

$$nAA + nBB = 2nAB, \quad n = 2, \ldots, z_1,$$

which can be a result of the rearrangement of atoms over the lattice sites, are ignored in the quasi-chemical approximation. The free energy of mixing of $A_x B_{1-x}$ regular solution per mole and its minimum condition in the quasi-chemical approximation as well as in the two-point approximation are:

$$f^M = \frac{1}{2}\alpha_{A-B}x_{AB} +$$
$$RT\left[-\frac{z_1}{2}\left(x - \frac{x_{AB}}{2}\right)\ln\left(x - \frac{x_{AB}}{2}\right)\right.$$
$$- z_1 x_{AB} \ln\frac{x_{AB}}{2} - \frac{z_1}{2}\left(1 - x - \frac{x_{AB}}{2}\right)\ln\left(1 - x - \frac{x_{AB}}{2}\right)$$
$$\left. + (z_1 - 1)x \ln x + (z_1 - 1)(1 - x)\ln(1 - x)\right]$$

and:

$$\frac{df^M}{dx_{AB}} = \alpha_{A-B} + \frac{z_1}{2}RT \ln \frac{\left(\frac{x_{AB}}{2}\right)^2}{\left(x - \frac{x_{AB}}{2}\right)\left(1 - x - \frac{x_{AB}}{2}\right)} = 0,$$

respectively. The concentration of AB pairs is expressed by:

$$x_{AB} = \frac{\sqrt{1 + 4(\eta^2 - 1)x(1 - x)} - 1}{\eta^2 - 1},$$

where $\eta = \exp\left\{\frac{\alpha_{A-B}}{z_1 RT}\right\}$.

4.4.2 Miscibility Gap

The free energy of mixing of $A_x B_{1-x}$ regular solution per mole is:

$$
\begin{aligned}
f^M = \frac{1}{2}\alpha_{A-B}x_{AB} + RT\Big[& -\frac{z_1}{2}\Big(x - \frac{x_{AB}}{2}\Big)\ln\Big(x - \frac{x_{AB}}{2}\Big) \\
& - z_1 x_{AB}\ln\frac{x_{AB}}{2} - \frac{z_1}{2}\Big(1 - x - \frac{x_{AB}}{2}\Big)\ln\Big(1 - x - \frac{x_{AB}}{2}\Big) \\
& + (z_1 - 1)x\ln x + (z_1 - 1)(1 - x)\ln(1 - x)\Big].
\end{aligned}
$$

$$(4.4.2.1)$$

The free energy of mixing in the two-point approximation also is a symmetrical function relatively to composition $x = \frac{1}{2}$. Therefore, we use all our arguments applied for the one-point approximation to describe the miscibility gap boundary. Correspondingly, the composition of the regular solution is an independent variable. The condition of the miscibility gap boundary of $A_x B_{1-x}$ regular solution in the two-point approximation as well as in the one-point approximation is:

$$\frac{df^M}{dx} = 0. \qquad (4.4.2.2)$$

After using the expression of the free energy of mixing (Eqn (4.4.2.1)), the condition for the miscibility gap boundary looks like this:

$$\frac{z_1}{2}\ln\frac{x - \frac{x_{AB}}{2}}{1 - x - \frac{x_{AB}}{2}} + (z_1 - 1)\ln\frac{1 - x}{x} = 0,$$

where the concentration of AB pairs x_{AB} corresponds to the minimum free energy of mixing. Thus, the miscibility gap boundary is described by the system of two equations:

$$\frac{df^M}{dx} = 0, \quad \frac{\partial f^M}{\partial x_{AB}} = 0 \text{ or}$$

$$\frac{z_1}{2}\ln\frac{x - \frac{1}{2}x_{AB}}{1 - x - \frac{1}{2}x_{AB}} + (z_1 - 1)\ln\frac{1 - x}{x} = 0,$$

$$\exp\left\{\frac{2\alpha_{A-B}}{z_1 RT}\right\} = \frac{\left(x - \frac{x_{AB}}{2}\right)\left(1 - x - \frac{x_{AB}}{2}\right)}{\left(\frac{x_{AB}}{2}\right)^2}.$$

The system is reduced to one equation given by:

$$\frac{z_1}{2}\ln\frac{1+2\left(\eta^2-1\right)x-\sqrt{1+4(\eta^2-1)x(1-x)}}{1+2(\eta^2-1)(1-x)-\sqrt{1+4(\eta^2-1)x(1-x)}}+(z_1-1)\ln\frac{1-x}{x}=0,$$

where $\eta = \exp\left\{\frac{\alpha_{A-B}}{z_1 RT}\right\}$.

4.5 TWO-POINT APPROXIMATION FOR TERNARY REGULAR SOLUTIONS

4.5.1 Helmholtz Free Energy and Short-Range Order

The basic clusters in the two-point approximation for ternary regular solutions are also the pairs of the nearest neighbor atoms (pairs of the nearest neighbor lattice points). There are six types of the pairs (AA, AB, AC, BB, BC, and CC) in $A_x B_y C_{1-x-y}$ regular solution. Atoms as an overlapping figure of the two basic clusters are subclusters in this approximation.

The types, the concentrations, and the numbers of the different configurations of the basic clusters and the subclusters are shown in Tables 4.5 and 4.6.

TABLE 4.5 Two-point clusters

Number	Type	Probability	Number of different configurations, $\alpha_l(2)$
1	AA	$x_1(2)$	1
2	AB	$x_2(2)$	2
3	AC	$x_3(2)$	2
4	BB	$x_4(2)$	1
5	BC	$x_5(2)$	2
6	CC	$x_6(2)$	1

TABLE 4.6 One-point clusters

Number	Type	Probability	Number of different configurations, $\alpha_l(1)$
1	A	$x_1(1)$	1
2	B	$x_2(1)$	1
3	C	$x_3(1)$	1

The Helmholtz free energy of mixing is given by $f^M = u^M - Ts^M$, where the internal energy of mixing is:

$$u^M = \alpha_{A-B}x_2(2) + \alpha_{A-C}x_3(2) + \alpha_{B-C}x_5(2)$$

$$= \alpha_{A-B}\frac{x_{AB}}{2} + \alpha_{A-C}\frac{x_{AC}}{2} + \alpha_{B-C}\frac{x_{BC}}{2}$$

The entropy coefficients of the two-point and one-point clusters are:

$$\delta(2) = -\frac{N(2)}{N} = -\frac{z_1}{2},$$

$$\delta(1) = -\frac{N(1)}{N} - M(1,2)\delta(2) = z_1 - 1.$$

The entropy of mixing is:

$$s^M = R\Big\{ -\frac{z_1}{2}\,[x_1(2)\ln x_1(2) + 2x_2(2)\ln x_2(2) + 2x_3(2)\ln x_3(2)$$

$$+ x_4(2)\ln x_4(2) + 2x_5(2)\ln x_5(2) + x_6(2)\ln x_6(2)]$$

$$+ (z_1 - 1)[x_1(1)\ln x_1(1) + x_2(1)\ln x_2(1) + x_3(1)\ln x_3(1)]\Big\},$$

where $x_1(2) = x_{AA} = x - \frac{x_{AB}}{2} - \frac{x_{AC}}{2}$, $x_2(2) = \frac{x_{AB}}{2}$, $x_3(2) = \frac{x_{AC}}{2}$,

$$x_4(2) = x_{BB} = y - \frac{x_{AB}}{2} - \frac{x_{BC}}{2}, \quad x_5(2) = \frac{x_{BC}}{2},$$

$$x_6(2) = x_{CC} = 1 - x - y - \frac{x_{AC}}{2} - \frac{x_{BC}}{2}.$$

If the composition of a regular solution expressed by $x_1(1) \equiv x$ and $x_2(1) \equiv y$ is fixed, then the free energy of mixing is a function of three independent variables—$x_2(2)$, $x_3(2)$, and $x_5(2)$—since $x_1(2) = x_1(1) - x_2(2) - x_3(2)$, $x_4(2) = x_2(1) - x_2(2) - x_3(2)$, $x_6(2) = 1 - x_1(1) - x_2(1) - x_3(2) - x_5(2)$, and $x_3(1) = 1 - x - y$. The free energy of mixing can be written as:

$$f^M = \frac{1}{2}(\alpha_{A-B}x_{AB} + \alpha_{A-C}x_{AC} + \alpha_{B-C}x_{BC})$$

$$- RT\Big\{ -\frac{z_1}{2}\Big[\Big(x - \frac{x_{AB}}{2} - \frac{x_{AC}}{2}\Big)\ln\Big(x - \frac{x_{AB}}{2} - \frac{x_{AC}}{2}\Big) + x_{AB}\ln\frac{x_{AB}}{2}$$

$$+ x_{AC}\ln\frac{x_{AC}}{2} + \Big(y - \frac{x_{AB}}{2} - \frac{x_{AC}}{2}\Big)\ln\Big(y - \frac{x_{AB}}{2} - \frac{x_{AC}}{2}\Big)$$

$$+ x_{BC}\ln\frac{x_{BC}}{2} + \Big(1 - x - y - \frac{x_{AC}}{2} - \frac{x_{BC}}{2}\Big)$$

$$\ln\Big(1 - x - y - \frac{x_{AC}}{2} - \frac{x_{BC}}{2}\Big)\Big]$$

$$+ (z_1 - 1)[x\ln x + y\ln y + (1 - x - y)\ln(1 - x - y)]\Big\}.$$

The concentrations of AB, AC, and BC pairs or short-range order are determined by the minimum free energy of mixing that is given by the system of equations:

$$\frac{\partial f^M}{\partial x_2(2)} = 0, \quad \frac{\partial f^M}{\partial x_3(2)} = 0, \quad \frac{\partial f^M}{\partial x_5(2)} = 0 \text{ or}$$

$$\alpha_{A-B} - z_1 RT\left[\frac{1}{2}\ln\left(x - \frac{x_{AB}}{2} - \frac{x_{AC}}{2}\right) + \ln\frac{x_{AB}}{2} + \frac{1}{2}\ln\left(y - \frac{x_{AB}}{2} - \frac{x_{BC}}{2}\right)\right] = 0,$$

$$\alpha_{A-C} - z_1 RT\left[\frac{1}{2}\ln\left(x - \frac{x_{AB}}{2} - \frac{x_{AC}}{2}\right) + \ln\frac{x_{AC}}{2}\right.$$
$$\left. + \frac{1}{2}\ln\left(1 - x - y - \frac{x_{AC}}{2} - \frac{x_{BC}}{2}\right)\right] = 0,$$

$$\alpha_{B-C} - z_1 RT\left[\frac{1}{2}\ln\left(y - \frac{x_{AB}}{2} - \frac{x_{BC}}{2}\right)\right.$$
$$\left. + \ln\frac{x_{BC}}{2} + \frac{1}{2}\ln\left(1 - x - y - \frac{x_{AC}}{2} - \frac{x_{BC}}{2}\right)\right] = 0.$$

4.5.2 Miscibility Gap

The miscibility gap boundary is derived by minimizing the free energy of mixing of the two-phase system represented as a canonical ensemble. The free energy of mixing is:

$$F^M = \sum_{i=1}^{2} F_i^M(N, N_A, N_B, N_i, N_{iA}, N_{iB}),$$

where $N = N_A + N_B + N_C$, N_A, N_B, and N_C are the total number and numbers of A, B, and C atoms that are considered as given values and N_i, N_{iA}, N_{iB} are the numbers of atoms in i-th phase that are the reciprocally dependent variables. There are three constraints between the numbers of atoms:

$$\varphi_1 = N_1 + N_2 - N = 0, \quad \varphi_2 = N_{1A} + N_{2A} - N_A = 0,$$
$$\varphi_3 = N_{1B} + N_{2B} - N_B = 0,$$

The minimum free energy of mixing obtained by using the method of the Lagrange undetermined multipliers can be calculated by the system of equations:

$$\frac{\partial L}{\partial N_i} = 0, \quad \frac{\partial L}{\partial N_{iA}} = 0, \quad \frac{\partial L}{\partial N_{iB}} = 0,$$

$$\frac{\partial F^M}{\partial N_{iAB}} = 0, \quad \frac{\partial F^M}{\partial N_{iAC}} = 0, \quad \frac{\partial F^M}{\partial N_{iBC}} = 0,$$

$$\varphi_j = 0, \quad (i = 1, 2; j = 1, ..., 3),$$

where $L = F^M + \sum_{j=1}^{3} \lambda_j \varphi_j$, λ_i are the Lagrange undetermined multipliers. The free energy of mixing of the i-th phase is:

$$
\begin{aligned}
F_i^M &= w_{A-B} N_{iAB} + w_{A-C} N_{iAC} + w_{B-C} N_{iBC} \\
&- k_B T \Bigg\{ -\frac{z_1}{2} \Bigg[\left(N_{iA} - \frac{N_{iAB} + N_{iAC}}{z_1} \right) \ln \left(\frac{N_{iA}}{N_i} - \frac{N_{iAB} + N_{iAC}}{z_1 N_i} \right) \\
&+ \left(N_{iB} - \frac{N_{iAB} + N_{iBC}}{z_1} \right) \ln \left(\frac{N_{iB}}{N_i} - \frac{N_{iAB} + N_{iBC}}{z_1 N_i} \right) \\
&+ \frac{2N_{iAB}}{z_1} \ln \frac{N_{iAB}}{z_1 N_i} + \frac{2N_{iAC}}{z_1} \ln \frac{N_{iAC}}{z_1 N_i} + \frac{2N_{iBC}}{z_1} \ln \frac{N_{iBC}}{z_1 N_i} \\
&+ \left(N_i - N_{iA} - N_{iB} - \frac{N_{iAC} - N_{iBC}}{z_1} \right) \\
&\ln \left(1 - \frac{N_{iA} - N_{iB}}{N_i} - \frac{N_{iAC} - N_{iBC}}{z_1 N_i} \right) \Bigg] \\
&+ (z_1 - 1) \Bigg[N_{iA} \ln \frac{N_{iA}}{N_i} + N_{iB} \ln \frac{N_{iB}}{N_i} \\
&+ (N_i - N_{iA} - N_{iB}) \ln \frac{N_i - N_{iA} - N_{iB}}{N_i} \Bigg] \Bigg\}.
\end{aligned}
$$

Finally, the miscibility gap boundary is described by the following system of equations:

$$\frac{z_1}{2} \ln \frac{1 - x_1 - y_1 - x_{1AC} - x_{1BC}}{1 - x_2 - y_2 - x_{2AC} - x_{2BC}} - (z_1 - 1) \ln \frac{1 - x_{1A} - x_{1B}}{1 - x_{2A} - x_{2B}} = 0,$$

$$
\begin{aligned}
&\frac{z_1}{2} \ln \frac{(x_1 - x_{1AB} - x_{1AC})(1 - x_2 - y_2 - x_{2AC} - x_{2BC})}{(x_2 - x_{2AB} - x_{2AC})(1 - x_1 - y_1 - x_{1AC} - x_{1BC})} \\
&- (z_1 - 1) \ln \frac{x_{1A}(1 - x_{2A} - y_{2A})}{x_{2A}(1 - x_{1A} - y_{1A})} = 0,
\end{aligned}
$$

$$
\begin{aligned}
&\frac{z_1}{2} \ln \frac{(y_1 - x_{1AB} - x_{1BC})(1 - x_2 - y_2 - x_{2AC} - x_{2BC})}{(y_2 - x_{2AB} - x_{2BC})(1 - x_1 - y_1 - x_{1AC} - x_{1BC})} \\
&- (z_1 - 1) \ln \frac{y_1(1 - x_2 - y_2)}{y_2(1 - x_1 - y_1)} = 0,
\end{aligned}
$$

$$
\begin{aligned}
&\alpha_{A-B} - z_1 RT \Bigg[\frac{1}{2} \ln \left(x_i - \frac{x_{iAB}}{2} - \frac{x_{iAC}}{2} \right) + \ln \frac{x_{iAB}}{2} \\
&+ \frac{1}{2} \ln \left(y_i - \frac{x_{iAB}}{2} - \frac{x_{iBC}}{2} \right) \Bigg] = 0,
\end{aligned}
$$

$$\alpha_{A-C} - z_1 RT \left[\frac{1}{2} \ln\left(x_i - \frac{x_{iAB}}{2} - \frac{x_{iAC}}{2} \right) + \ln \frac{x_{iAC}}{2} \right.$$
$$\left. + \frac{1}{2} \ln\left(1 - x_i - y_i - \frac{x_{iAC}}{2} - \frac{x_{iBC}}{2} \right) \right] = 0,$$

$$\alpha_{B-C} - z_1 RT \left[\frac{1}{2} \ln\left(y_i - \frac{x_{iAB}}{2} - \frac{x_{iBC}}{2} \right) + \ln \frac{x_{iBC}}{2} \right.$$
$$\left. + \frac{1}{2} \ln\left(1 - x_i - y_i - \frac{x_{iAC}}{2} - \frac{x_{iBC}}{2} \right) \right] = 0$$

where $x_2 = \frac{x - \gamma x_1}{1 - \gamma}$, $y_2 = \frac{y - \gamma y_1}{1 - \gamma}$, $\gamma = \frac{N_1}{N}$, and $x_1, y_1, \gamma = \frac{N_1}{N}$, x_{iAB}, x_{iAC}, and x_{iBC}
($i = 1, 2$) are the independent variables.

4.6 THREE-POINT APPROXIMATION FOR BINARY REGULAR SOLUTION WITH TRIANGULAR LATTICE

4.6.1 Helmholtz Free Energy and Short-Range Order

Triads of atoms (triangles) are chosen as basic clusters in the three-point approximation for a binary regular solution with the triangle lattice. There are four types of triads of atoms (AAA, AAB, ABB, and BBB) in the $A_x B_{1-x}$ regular solution. The triads AAB as well as triads ABB can be disposed on the same lattice sites by three configurationally different variants as triads AAB, ABA, BAA, and triads ABB, BAB, BBA, respectively, or the number of different configurations that can be generated by the symmetry operations of the basic clusters AAB and ABB is equal to three. The pairs of atoms (bonds) and atoms are the subclusters, since they are the overlapping figures of the basic clusters.

The types, the concentrations, and the numbers of the different configurations of the basic clusters and the subclusters are shown in Tables 4.7–4.9.

The internal energy of mixing is expressed by:

$$u^M = \alpha_{A-B} x_2(2) = \frac{1}{2} \alpha_{A-B} x_{AB}.$$

The entropy coefficients for the three-, two-, and one-point clusters are given, respectively, by:

$$\delta(3) = -\frac{N(3)}{N} = -2,$$

$$\delta(2) = -\frac{N(2)}{N} - M(2,3)\delta(3) = 3,$$

$$\delta(1) = -\frac{N(1)}{N} - M(1,3)\delta(3) - M(1,2)\delta(2) = -1.$$

TABLE 4.7   Three-point clusters

Number	Type	Concentration	Number of different configurations, $\alpha_i(3)$
1	AAA	$x_1(3)$	1
2	AAB	$x_2(3)$	3
3	ABB	$x_3(3)$	3
4	BBB	$x_4(3)$	1

TABLE 4.8   Two-point clusters

Number	Type	Concentration	Number of different configurations, $\alpha_i(2)$
1	AA	$x_1(2)$	1
2	AB	$x_2(2)$	2
3	BB	$x_3(2)$	1

TABLE 4.9   One-point clusters

Number	Type	Concentration	Number of different configurations, $\alpha_2(1)$
1	A	$x_1(1)$	1
2	B	$x_2(1)$	1

Accordingly, the entropy of mixing is:

$$s^M = R\left[-2\sum_{i=1}^{4}\alpha_i(3)x_i(3)\ln x_i(3) + 3\sum_{j=1}^{3}\alpha_j(2)x_j(2)\ln x_j(2) \right.$$
$$\left. - \sum_{k=1}^{2}x_k(1)\ln x_k(1)\right]$$

The concentrations of the two-point subclusters are represented as functions of the concentrations of basic clusters, which are the variables:

$$x_1(2) = x_1(3) + x_2(3), \quad x_2(2) = 2x_2(3) + 2x_3(3), \quad x_3(2) = x_3(3) + x_4(3).$$

There are four reciprocally dependent variables $x_i(3)$, $(i = 1,...,4)$ since there are two constraints between them if the composition of the regular solution $x \equiv x_1(1)$ is considered as a given value:

$$x_1(3) + 3x_2(3) + 3x_3(3) + x_4(3) - 1 = 0,$$
$$x_1(3) + 2x_2(3) + x_3(3) - x_1(1) = 0.$$

Therefore, there are two independent variables. Let $x_3(3)$ and $x_4(3)$ be the independent variables. In such a case, the concentrations of the other basic clusters and the subclusters are:

$$x_1(3) = -2 + 3x + 3x_3(3) + 2x_4(3), \quad x_2(3) = 1 - x - 2x_3(3) - x_4(3),$$
$$x_1(2) = -1 + 2x + x_3(3) + x_4(3),$$
$$x_2(2) = 1 - x - x_3(3) - x_4(3), \quad x_3(2) = x_3(3) + x_4(3).$$

The free energy of mixing is expressed by:

$$
\begin{aligned}
f^M = {}& \alpha_{A-B}[1 - x - x_3(3) - x_4(3)] \\
& - RT\{ - 2[- 2 + 3x + 3x_3(3) + 2x_4(3)] \\
& \quad \ln[- 2 + 3x + 3x_3(3) + 2x_4(3)] - 6[1 - x - 6x_3(3) - x_4(3)] \\
& \quad \ln[1 - x - 6x_3(3) - x_4(3)] - 6x_3(3)\ln x_3(3) - 2x_4(3)\ln x_4(3) \\
& \quad + 3[- 1 + 2x + x_3(3) + x_4(3)]\ln[- 1 + 2x + x_3(3) + x_4(3)] \\
& \quad + 6[1 - x - x_3(3) - x_4(3)]\ln[1 - x - x_3(3) - x_4(3)] \\
& \quad + 3[x_3(3) + x_4(3)]\ln[x_3(3) + x_4(3)] \\
& \quad - x \ln x - (1 - x)\ln(1 - x)\}.
\end{aligned}
$$

The short-range order:

$$X_{AB} = 2x_2(2) = 2[1 - x - x_3(3) - x_4(3)]$$

for the given composition is obtained by minimizing the free energy of mixing:

$$\frac{\partial f^M}{dx_3(3)} = 0 \text{ and } \frac{\partial f^M}{dx_4(3)} = 0$$

which can be rewritten as:

$$\frac{[1 - x - 2x_3(3) - x_4(3)]^3 x_4(3)}{[- 2 + 3x + 3x_3(3) + 3x_4(3)][x_3(3)]^3} - 1 = 0 \text{ and}$$

$$\frac{[1 - x - 2x_3(3) - 2x_4(3)]^4[- 1 + 2x + x_3(3) + x_4(3)]^3[x_3(3) + x_4(3)]^3}{[- 2 + 3x + 3x_3 + 2x_4(3)]^4[x_4(3)]^2[1 - x - x_3(3) - x_4(3)]^6} - \eta^6 = 0,$$

where $\eta = \exp\left\{\frac{\alpha_{A-B}}{z_1 RT}\right\}$.

If composition is $x = \frac{1}{2}$, then there are two additional constraints that are given by $x_1(3) = x_4(3)$ and $x_2(3) = x_3(3)$. Moreover, the first two constraints reduce to one constraint:

$$6x_3(3) + 2x_4(3) - 1 = 0.$$

Thus, there is one independent variable, $x_3(3)$, and the free energy of mixing is:

$$f^M = 2\alpha_{A-B}x_3(3) - RT\left\{-4\left[\frac{1}{2} - 3x_3(3)\right]\ln\left[\frac{1}{2} - 3x_3(3)\right] - 12x_3(3)\ln x_3(3)\right.$$
$$+ 6\left[\frac{1}{2} - 2x_3(3)\right]\ln\left[\frac{1}{2} - 2x_3(3)\right]$$
$$\left. + 12x_3(3)\ln[2x_3(3)] - \ln\frac{1}{2}\right\}.$$

The short-range order $x_{AB} = 2x_2(2) = 4x_3(3)$ at composition $x = \frac{1}{2}$ is also obtained by minimizing the free energy of mixing, which is written as $\frac{\partial f^M}{\partial x_3(3)} = 0$ or $x_{AB} = \frac{2-\eta}{3-\eta}$.

4.6.2 Miscibility Gap

The free energy of mixing given by:

$$f^M = \alpha_{A-B}[1 - x - x_3(3) - x_4(3)]$$
$$- RT\{- 2[-2 + 3x + 3x_3(3) + 2x_4(3)]$$
$$\ln[-2 + 3x + 3x_3(3) + 2x_4(3)] - 6[1 - x - 6x_3(3) - x_4(3)]$$
$$\ln[1 - x - 6x_3(3) - x_4(3)] - 6x_3(3)\ln x_3(3) - 2x_4(3)\ln x_4(3)$$
$$+ 3[-1 + 2x + x_3(3) + x_4(3)]\ln[-1 + 2x + x_3(3) + x_4(3)]$$
$$+ 6[1 - x - x_3(3) - x_4(3)]\ln[1 - x - x_3(3) - x_4(3)]$$
$$+ 3[x_3(3) + x_4(3)]\ln[x_3(3) + x_4(3)] - x\ln x - (1 - x)\ln(1 - x)\}$$

is the symmetric function relatively to composition $x = \frac{1}{2}$. Therefore, the miscibility gap temperature is described by the system of three equations:

$$\frac{df^M}{dx} = 0, \quad \frac{\partial f^M}{\partial x_3(3)} = 0 \text{ and } \frac{\partial f^M}{\partial x_4(3)} = 0,$$

since, in such a case, the free energy of mixing is the function of three independent variables x, $x_3(3)$ and $x_4(3)$ as far as the composition is not a fixed value. This system of equations can be rewritten as:

$$\frac{[1 - x - 2x_3(3) - x_4(3)]^3 x_4(3)}{[-2 + 3x + 3x_3(3) + 2x_4(3)][x_3(3)]^3} - 1 = 0,$$

$$\frac{[-1+2x+x_3(3)+x_4(3)]^3(1-x)[x_4(3)]^2}{[-2+3x+3x_3(3)+2x_4(3)]^2 x[x_3(3)+x_4(3)]^3} - 1 = 0,$$

$$\frac{[-2+3x+3x_3(3)+2x_4(3)]^2[x_3(3)]^2[1-x-x_3(3)-x_4(3)]^2}{[1-x-2x_3(3)-x_4(3)]^4[-1+2x+x_3(3)+x_4(3)][x_3(3)+x_4(3)]} - \eta^2 = 0,$$

where $\eta = \exp\left\{\frac{\alpha_{A-B}}{z_1 RT}\right\}$. If composition is $x = \frac{1}{2}$, the miscibility gap temperature is $T^{MG} = 0.326\frac{\alpha_{A-B}}{R}$ and the concentration of AB pairs at this temperature is $x_{AB} = \frac{1}{4}$.

The exact solution for the Ising ferromagnet on the triangular lattice in the zero external magnetic field was obtained for the first time by Wannie [12]. The Ising ferromagnet in the zero external magnetic field is equivalent to $A_{0.5}B_{0.5}$ regular solution. The extension of the results obtained by Wannie to $A_{0.5}B_{0.5}$ regular solution leads to the following expressions for the Helmholtz free energy of mixing, the internal energy of mixing, and the concentration of AB pairs given, respectively, by:

$$f^M = -RT \ln\left(e^{3L}+e^{-L}\right) - \frac{RT}{8\pi^2}\int_0^{2\pi}\int_0^{2\pi}\ln\left(1+4\kappa\cos x\cos y - 4\kappa\cos^2 y\right)dxdy,$$

$$u = -\frac{\alpha_{A-B}}{6(1-\mu)}\left\{1 - \frac{\mu(3-\mu)}{1+\mu}\times\frac{\frac{2}{\pi}K(k)}{\sqrt{1+\frac{1}{4}(1-\mu)^2}}\right\},$$

$$x_{AB}\left(T \geq T^{MG}\right) = \frac{1}{2}+\frac{1}{3(1-\mu)}\left[1 - \frac{\mu(3-\mu)}{1+\mu}\frac{2K(k)}{\pi\sqrt{1+\frac{(1-\mu)^2}{4}}}\right],$$

where $L = \frac{\alpha_{A-B}}{2z_1 RT}$, $\kappa \equiv \frac{e^{4L}-1}{(e^{4L}+1)^2}$, $\mu = 1 - 2\tanh\frac{\alpha_{A-B}}{z_1 RT}$, $k = \frac{1-\mu}{1+\mu}\sqrt{\frac{4+(1+\mu)^2}{4+(1-\mu)^2}}$, and

$$K(k) = \int_0^{\pi/2}\frac{d\varphi}{\sqrt{1-k^2\sin^2\varphi}}$$ is Legendre's complete elliptic integral of the first kind. The miscibility gap temperature and concentration of AB pairs at the miscibility gap temperature are given, respectively, by $T^{MG} = 0.304\frac{\alpha_{A-B}}{R}$ and $x_{AB}(T^{MG}) = \frac{1}{6}$.

The concentration dependencies of the miscibility gap temperatures obtained by using the one-, two-, and three-point approximations as well as obtained by using the exact solution at composition $x = \frac{1}{2}$ are shown in Figure 4.2. The miscibility gap temperatures differ significantly in the

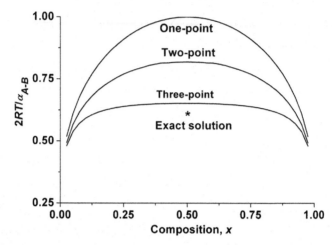

FIGURE 4.2 The miscibility gap temperatures obtained by using the one-, two- and three-point approximations and the exact solution.

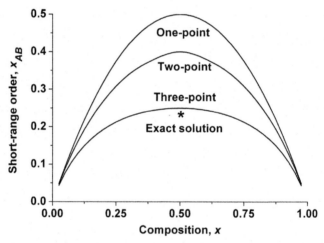

FIGURE 4.3 The AB pair concentrations at the miscibility gap temperature obtained by using the one-, two- and three-point approximations and exact solution.

wide region of the concentrations and decrease with the increase of the accuracy of the approximation.

The AB pair concentration dependencies on the composition at the miscibility gap temperature obtained by using the one-, two-, and three-point approximations as well as the exact solution at composition $x = \frac{1}{2}$ are shown in Figure 4.3.

The AB pair concentration dependencies on temperature for $A_{0.5}B_{0.5}$ regular solution estimated by using the one-, two-, and three-point approximations and exact solution are shown in Figure 4.4. The results

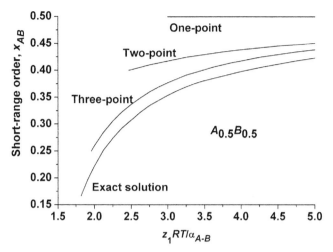

FIGURE 4.4 The temperature dependencies of AB pair concentrations for $A_{0.5}B_{0.5}$ regular solution estimated by using the one-, two- and three-point approximations and exact solution.

obtained by using the three-point approximation and by the exact solution demonstrate a strong preference for the formation of AA and BB pairs.

Moreover, the curves shown in Figures 4.2–4.4 demonstrate the significant increase of the accuracy of the results with the increase of the basic cluster.

4.7 FOUR-POINT APPROXIMATION FOR BINARY REGULAR SOLUTION WITH SIMPLE SQUARE LATTICE

4.7.1 Helmholtz Free Energy and Short-Range Order

For the first time, the four-point (square) approximation for the Ising ferromagnet with the simple square lattice in the zero external magnetic field was studied in the fundamental work of Kikuchi [2]. Pairs of spins and spins are used as subclusters, since they are the overlapping figures of the basic clusters. All results obtained by Kikuchi can be extended on $A_{0.5}B_{0.5}$ regular solution.

The four-point approximation for A_xB_{1-x} ($0 < x < 1$) regular solution on the simple square lattice will be represented here by using Baker's approach. The quadruple (square) of atoms is the basic cluster. The pair of atoms (bond) and atom are the overlapping figures of the basic clusters and, therefore, they are subclusters.

The types, the concentrations, and the numbers of the different configurations of the basic clusters and the subclusters are shown in Tables 4.10–4.12.

TABLE 4.10 Four-point clusters

Number	Type	Concentration	Number of different configurations, $\alpha_l(4)$
1	AAAA	$x_1(4)$	1
2	AAAB	$x_2(4)$	4
3	AABB	$x_3(4)$	4
4	ABAB	$x_4(4)$	2
5	ABBB	$x_5(4)$	4
6	BBBB	$x_6(4)$	1

TABLE 4.11 Two-point clusters

Number	Type	Concentration	Number of different configurations, $\alpha_l(2)$
1	AA	$x_1(2)$	1
2	AB	$x_2(2)$	2
3	BB	$x_3(2)$	1

TABLE 4.12 One-point clusters

Number	Type	Concentration	Number of different configurations, $\alpha_l(1)$
1	A	$x_1(1)$	1
2	B	$x_2(1)$	1

The entropy coefficients for the four-, two-, and one-point clusters are:

$$\delta(4) = -\frac{N(4)}{N} = -1,$$

$$\delta(2) = -\frac{N(2)}{N} - M(2,4)\delta(4) = 2,$$

$$\delta(1) = -\frac{N(1)}{N} - M(1,4)\delta(4) - M(1,2)\delta(2) = -1.$$

The free energy of mixing is written as:

$$f^M = \alpha_{A-B} x_2(2) - RT \left[-\sum_{i=1}^{6} \alpha_i(4) x_i(4) \ln x_i(4) + 2 \sum_{j=1}^{3} \alpha_j x_j(2) \ln x_j(2) \right.$$

$$\left. - \sum_{k=1}^{2} \alpha_k x_k(1) \ln x_k(1) \right].$$

The free energy of mixing $f^M = u^M - Ts^M$ is the symmetric function relatively to composition $x = \frac{1}{2}$ as well as the free energies of mixing of $A_x B_{1-x}$ regular solutions derived in the one- and two-point approximations. The concentrations of the basic clusters are considered as variables. The concentrations of the pairs are expressed by the concentrations of the basic clusters as:

$$x_1(2) = x_1(4) + 2x_2(4) + x_3(4),$$
$$x_2(2) = x_2(4) + x_3(4) + x_4(4) + x_5(4),$$
$$x_3(2) = x_3(4) + 2x_5(4) + x_6(4).$$

If the composition of the regular solution is a fixed value, then the concentrations of the four-point clusters are connected by two constraints:

$$x_1(4) + 4x_2(4) + 4x_3(4) + 2x_4(4) + 4x_5(4) + x_6(4) = 1,$$
$$x_1(4) + 3x_2(4) + 2x_3(4) + x_4(4) + x_5(4) = x. \qquad (4.7.1.1)$$

Thus, there are four independent variables. Let $x_1(4)$, $x_2(4)$, $x_3(4)$, and $x_4(4)$ be the independent variables. In such a case, the other concentrations of the four-point clusters are:

$$x_5(4) = x - x_1(4) - 3x_2(4) - 2x_3(4) - x_4(4),$$
$$x_6(4) = 1 - 4x + 3x_1(4) + 8x_2(4) + 4x_3(4) + 2x_4(4).$$

The short-range order for the given value of the composition is derived by minimizing the free energy of mixing or the system of four equations:

$$\frac{\partial f^M}{\partial x_1(4)} = 0, \quad \frac{\partial f^M}{\partial x_2(4)} = 0, \quad \frac{\partial f^M}{\partial x_3(4)} = 0, \quad \frac{\partial f^M}{\partial x_4(4)} = 0.$$

If the composition is $x = \frac{1}{2}$, then the number of the independent variables reduces to three, since in such a case there are two additional constraints: $x_1(4) = x_6(4)$ and $x_2(4) = x_5(4)$, and two constraints (Eqn (4.7.1.1)) reduce to one.

The exact values of the concentrations of AB pairs in $A_{0.5}B_{0.5}$ regular solution as a function of the temperature can be deduced from the Onzager solution for the Ising ferromagnet in the zero external magnetic field [13] as:

$$x_{AB} = \frac{1}{2} - \frac{1}{4}\coth\frac{\alpha_{A-B}}{z_1 RT}\left[1 + \frac{2}{\pi}\chi K(k)\right],$$

where $\chi = 2\left(\tanh\frac{\alpha_{A-B}}{z_1 RT}\right)^2 - 1$, $K(k) = \int_0^{\pi/2}\frac{d\varphi}{\sqrt{1 - k^2\sin^2\varphi}}$ is Legendre's

complete integral of the first kind, and $k = 2\dfrac{\sinh\frac{\alpha_{A-B}}{z_1 RT}}{\left(\cosh\frac{\alpha_{A-B}}{z_1 RT}\right)^2}$.

4.7.2 Miscibility Gap

The free energy of mixing given by:

$$f^M = \alpha_{A-B}x_2(2) - RT\left[-\sum_{i=1}^{6}\alpha_i(4)x_i(4)\ln x_i(4) + 2\sum_{j=1}^{3}\alpha_j(2)x_j(2)\ln x_j(2)\right.$$
$$\left. - x\ln x - (1-x)\ln(1-x)\right]$$

is the symmetric function relatively to composition $x = \frac{1}{2}$. Therefore, the miscibility gap temperature is described by the system of five equations:

$$\frac{df^M}{dx} = 0, \quad \frac{\partial f^M}{\partial x_1(4)} = 0, \quad \frac{\partial f^M}{\partial x_2(4)} = 0, \quad \frac{\partial f^M}{\partial x_3(4)} = 0 \text{ and } \frac{\partial f^M}{\partial x_4(4)} = 0$$

since the composition of A_xB_{1-x} regular solution is considered as an independent variable. If composition is $x = \frac{1}{2}$, then the number of the independent variables reduces to three, since, in such a case, there are two additional constraints: $x_1(4) = x_6(4)$, $x_2(4) = x_5(4)$, and two constraints:

$$x_1(4) + 4x_2(4) + 4x_3(4) + 2x_4(4) + 4x_5(4) + x_6(4) = 1,$$
$$x_1(4) + 3x_2(4) + 2x_3(4) + x_4(4) + x_5(4) = x$$

reduce to one constraint, given by:

$$x_1(4) + 4x_2(4) + 2x_3(4) + x_4(4) = \frac{1}{2}.$$

For composition $x = \frac{1}{2}$, the miscibility gap temperatures and the concentrations of AB pairs in the one-, two-, and four-point approximations and obtained by using the exact solution are given by:

$$\text{one-point approximation: } T^{MG} = \frac{\alpha_{A-B}}{2R}, \quad x_{AB} = \frac{1}{2},$$

$$\text{two-point approximation: } T^{MG} = 0.721 \frac{\alpha_{A-B}}{2R}, \quad x_{AB} = \frac{1}{3},$$

$$\text{four-point approximation: } T^{MG} = 0.607 \frac{\alpha_{A-B}}{2R}, \quad x_{AB} = 0.219,$$

$$\text{exact solution: } T^{MG} = 0.568 \frac{\alpha_{A-B}}{2R}, \quad x_{AB} = 0.146.$$

The short-range order x_{AB} estimated by using the one-, two-, and four-point approximations and the exact solution at the temperatures higher than the miscibility gap temperatures and at composition $x = \frac{1}{2}$ are shown in Figure 4.5.

These results obtained by using the four-point approximation and the exact solution demonstrate the strong preference formation of AA and BB pairs over AB pairs. The curves shown in Figure 4.5 demonstrate the significant increase of the accuracy of the results with the increase of the basic cluster.

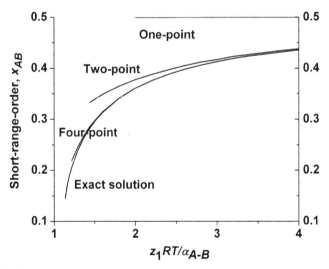

FIGURE 4.5 The short-range order for the $A_{0.5}B_{0.5}$ regular solution estimated by using the one-, two- and four-point approximations and the exact solution at composition $x = \frac{1}{2}$.

4.8 FOUR-POINT APPROXIMATION FOR BINARY REGULAR SOLUTIONS WITH FACE-CENTERED CUBIC AND HEXAGONAL CLOSE-PACKED LATTICES

4.8.1 Helmholtz Free Energy and Short-Range Order

The Ising ferromagnet with the face-centered cubic lattice in the zero external magnetic field is the lattice model most studied by the cluster variation method. The different clusters were used as basic clusters. Kikuchi was the first to consider the quadruple (tetrahedron) of spins as the basic cluster in his pioneering work [2]. The pairs of spins (bonds) and spins (lattice points) were subclusters since they are overlapping figures of the basic clusters. The hexagonal close-packed lattice consists of the same numbers of tetrahedrons and pairs of spins as the face-centered cubic lattice. Moreover, pairs and lattice points also are overlapping figures of tetrahedrons in this lattice. Therefore, the entropy of mixing as well as the internal energy of mixing of the Ising ferromagnet with the face-centered cubic and hexagonal close-packed lattices derived by the four-point approximation, should be equal to one another. $A_x B_{1-x}$ regular solutions with the face-centered cubic and hexagonal close-packed lattices are considered here by using Baker's approach.

The types, the concentrations, and the numbers of the different configurations of the basic clusters and subclusters are shown in Tables 4.13–4.15.

The entropy coefficients for the four-, two-, and one-point clusters are given by:

$$\delta(4) = -\frac{N(4)}{N} = -2,$$

$$\delta(2) = -\frac{N(2)}{N} - M(2,4)\delta(4) = 6,$$

TABLE 4.13   Four-point clusters

Number	Type	Concentration	Number of different configurations, $\alpha_l(4)$
1	AAAA	$x_1(4)$	1
2	AAAB	$x_2(4)$	4
3	AABB	$x_3(4)$	6
4	ABBB	$x_4(4)$	4
5	BBBB	$x_5(4)$	1

TABLE 4.14 Two-point clusters

Number	Type	Concentration	Number of different configurations, $\alpha_l(2)$
1	AA	$x_1(2)$	1
2	AB	$x_2(2)$	2
3	BB	$x_3(2)$	1

TABLE 4.15 One-point clusters

Number	Type	Concentration	Number of different configurations, $\alpha_l(1)$
1	A	$x_1(1)$	1
2	B	$x_2(1)$	1

$$\delta(1) = -\frac{N(1)}{N} - M(1,4)\delta(4) - M(1,2)\delta(2) = -5.$$

The free energy of mixing looks like this:

$$f^M = \alpha_{A-B}x_2(2) - RT\left[-2\sum_{i=1}^{5}\alpha_i(4)x_i(4)\ln x_i(4) + 6\sum_{j=1}^{3}\alpha_j(2)x_j(2)\ln x_j(2)\right.$$
$$\left. - 5\sum_{k=1}^{2}\alpha_k(1)x_k(1)\ln x_k(1)\right].$$

The free energy of mixing $f^M = u^M - Ts^M$ is the symmetric function relatively to composition $x = \frac{1}{2}$ as well as the free energies of mixing represented in the one- and two-point approximations. The concentrations of the basic clusters are used as variables. The concentrations of pairs are expressed by the concentrations of the tetrahedrons as:

$$x_1(2) = x_1(4) + 2x_2(4) + x_3(4),$$
$$x_2(2) = x_2(4) + 2x_3(4) + x_4(4),$$
$$x_3(2) = x_3(4) + 2x_4(4) + x_5(4).$$

The concentrations of the tetrahedrons are connected by two constraints for the given composition written as:

$$x_1(4) + 4x_2(4) + 6x_3(4) + 4x_4(4) + x_5(4) = 1,$$
$$x_1(4) + 3x_2(4) + 3x_3(4) + x_4(4) = x.$$

Thus, there are three independent variables. Let $x_1(4)$, $x_2(4)$, and $x_3(4)$ be the independent variables. In such a case, other concentrations of the tetrahedrons are:

$$x_4(4) = x - x_1(4) - 3x_2(4) - 3x_3(4),$$

$$x_5(4) = 1 - 4x + 3x_1(4) + 8x_2(4) + 6x_3(4)$$

The short-range order is derived by minimizing the free energy of mixing that is given by the system of three equations:

$$\frac{\partial f^M}{\partial x_1(4)} = 0, \quad \frac{\partial f^M}{\partial x_2(4)} = 0, \quad \frac{\partial f^M}{\partial x_3(4)} = 0.$$

If composition is $x = \frac{1}{2}$, then the number of the independent variables reduces to two, since there are two additional constraints, $x_1(4) = x_5(4)$ and $x_2(4) = x_4(4)$, and two constraints:

$$x_4(4) = x - x_1(4) - 3x_2(4) - 3x_3(4),$$

$$x_5(4) = 1 - 4x + 3x_1(4) + 8x_2(4) + 6x_3(4)$$

reduce to one constraint. Accordingly, the concentration of AB pairs is obtained by the system of equations [2]:

$$x_{AB} = \frac{2}{\varphi^3 - \varphi^2 + \varphi + 3},$$

$$\frac{2\varphi^2}{\varphi^3 - \varphi^2 + \varphi + 1} = \eta,$$

where $\eta = \exp\left\{\frac{\alpha_{A-B}}{z_1 RT}\right\}$ and z_1 is the coordination number of the nearest neighbors.

4.8.2 Miscibility Gap

The free energy of mixing of $A_x B_{1-x}$ solution is:

$$f^M = \alpha_{A-B} x_2(2) - RT \left[-2 \sum_{i=1}^{5} \alpha_i(4) x_i(4) \ln x_i(4) + 6 \sum_{j=1}^{3} \alpha_j(2) x_j(2) \ln x_j(2) \right.$$

$$\left. - 5 \sum_{k=1}^{2} \alpha_k(1) x_k(1) \ln x_k(1) \right].$$

The miscibility gap is determined by the minimization of the free energy of mixing of the two-phase system. The free energy of mixing is the symmetric function of the composition x with respect to the point $x = \frac{1}{2}$, i.e., $f^M(x) = f^M(1 - x)$. Accordingly, the miscibility gap boundary is derived by the system of equations that is given by:

$$\frac{df^M}{dx} = 0, \quad \frac{\partial f^M}{\partial x_1(4)} = 0, \quad \frac{\partial f^M}{\partial x_2(4)} = 0, \quad \frac{\partial f^M}{\partial x_3(4)} = 0.$$

The miscibility gap temperatures can be interpolated with the fourth-order polynomial as:

$$T^{MG}(4) = (- 4.6103371x^4 + 9.22073x^3 - 7.193803x^2 + 2.583084x$$
$$+ 0.4777) \frac{\alpha_{A-B}}{2R},$$

where $0.05 \leq x \leq 0.95$. The miscibility gap temperature at composition $x = \frac{1}{2}$ is written as $T^{MG}(4) = 0.835 \frac{\alpha_{A-B}}{2R}$. For comparison, the miscibility gap temperatures obtained by the two-point and one-point approximations at composition $x = \frac{1}{2}$ are equal, respectively, to $T^{MG}(2) = 0.913 \frac{\alpha_{A-B}}{2R}$ and $T^{MG}(1) = \frac{\alpha_{A-B}}{2R}$. The short-range order x_{AB} estimated by using the one-, two- and four-point approximations at temperatures higher than the miscibility gap temperatures at composition $x = \frac{1}{2}$ are shown in Figure 4.6.

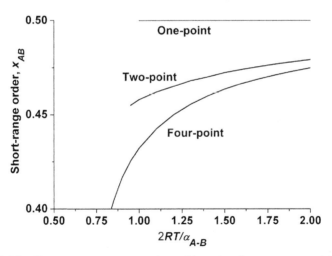

FIGURE 4.6 The concentrations x_{AB} estimated by using the one-, two- and four-point approximations at composition $x = \frac{1}{2}$.

4.9 SIX-POINT APPROXIMATION FOR BINARY REGULAR SOLUTION WITH DIAMOND LATTICE

4.9.1 Helmholtz Free Energy and Short-Range Order

The hexads of atoms, or the closed chain consisting of six atoms and six bonds between them, are considered as basic clusters for A_xB_{1-x} regular solution with the diamond lattice. There are 13 types of hexads. The triads (angles) of atoms, the pairs of atoms, and the atoms are the overlapping figures of hexads and, therefore, are the subclusters.

The types, the concentrations and the numbers of the different configurations of the basic clusters and the subclusters are shown in Tables 4.16–4.19.

The different configurations of the *BBABAA* hexad related by the symmetry operations of the diamond lattice are shown in Figure 4.7.

The entropy coefficients for the hexads, the triads, the pairs, and the atoms are written, respectively, as:

$$\delta(6) = -\frac{N(6)}{N} = -2,$$

$$\delta(3) = -\frac{N(3)}{N} - M(3,6)\delta(6) = 6,$$

TABLE 4.16 Six-point clusters

Number	Type	Concentration	Number of different configurations, $\alpha_l(6)$
1	*AAAAAA*	$x_1(6)$	1
2	*BAAAAA*	$x_2(6)$	6
3	*BBAAAA*	$x_3(6)$	6
4	*BABAAA*	$x_4(6)$	6
5	*BAABAA*	$x_5(6)$	3
6	*BBBAAA*	$x_6(6)$	6
7	*BBABAA*	$x_7(6)$	12
8	*BABABA*	$x_8(6)$	2
9	*ABBABB*	$x_9(6)$	3
10	*ABABBB*	$x_{10}(6)$	6
11	*AABBBB*	$x_{11}(6)$	6
12	*ABBBBB*	$x_{12}(6)$	6
13	*BBBBBB*	$x_{13}(6)$	1

TABLE 4.17 Three-point clusters

Number	Type	Concentration	Number of different configurations, $\alpha_l(3)$
1	AAA	$x_1(3)$	1
2	AAB	$x_2(3)$	2
3	ABA	$x_3(3)$	1
4	BAB	$x_4(3)$	1
5	BBA	$x_5(3)$	2
6	BBB	$x_6(3)$	1

TABLE 4.18 Two-point clusters

Number	Type	Concentration	Number of different configurations, $\alpha_l(2)$
1	AA	$x_1(2)$	1
2	AB	$x_2(2)$	2
3	BB	$x_3(2)$	1

TABLE 4.19 One-point clusters

Number	Type	Concentration	Number of different configurations, $\alpha_2(1)$
1	A	$x_1(1)$	1
2	B	$x_2(1)$	1

$$\delta(2) = -\frac{N(2)}{N} - M(2,6)\delta(6) - M(2,3)\delta(3) = -2,$$

$$\delta(1) = -\frac{N(1)}{N} - M(1,6)\delta(6) - M(1,3)\delta(3) - M(1,2)\delta(2) = -3.$$

Thus, the entropy looks like this:

$$s^M = R\left[-2\sum_{i=1}^{13}\alpha_i(6)x_i(6)\ln x_i(6) + 6\sum_{j=1}^{6}\alpha_j(3)x_j(3)\ln x_j(3) \right.$$
$$\left. -2\sum_{k=1}^{3}\alpha_k(2)x_k(2)\ln x_k(2) - 3\sum_{l=1}^{2}\alpha_l(1)x_l(1)\ln x_l(1) \right]$$

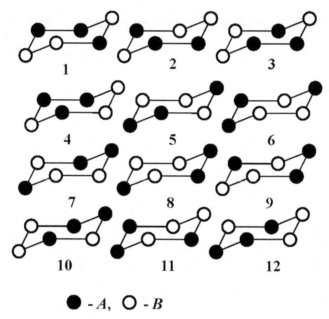

$$\bullet - A, \bigcirc - B$$

FIGURE 4.7   The different configurations of $BBABAA$ hexad.

The internal energy of mixing is $u^M = \alpha_{A-B} x_2(2)$.

The concentrations of the basic clusters are variables. The concentrations of the pairs are expressed by:

$$x_1(2) = x_1(6) + 4x_2(6) + 3x_3(6) + 2x_4(6) + x_5(6) + 2x_6(6) + 2x_7(6)$$
$$+ x_{11}(6),$$

$$x_2(2) = x_2(6) + x_3(6) + 2x_4(6) + x_5(6) + x_6(6) + 4x_7(6) + x_8(6) + x_9(6)$$
$$+ 2x_{10}(6) + x_{11}(6) + x_{12}(6),$$

$$x_3(2) = x_3(6) + 2x_6(6) + 2x_7(6) + x_9(6) + 2x_{10}(6) + 3x_{11}(6) + 4x_{12}(6)$$
$$+ x_{13}(6).$$

The concentrations of the triads are written as:

$$x_1(3) = x_1(6) + 3x_2(6) + 2x_3(6) + x_4(6) + x_6(6),$$
$$x_2(3) = x_2(6) + x_3(6) + 2x_4(6) + x_5(6) + x_6(6) + x_7(6) + x_{11}(6),$$
$$x_3(3) = x_2(6) + 2x_4(6) + x_5(6) + 2x_7(6) + x_8(6) + x_{10}(6),$$
$$x_4(3) = x_4(6) + 2x_7(6) + x_8(6) + x_9(6) + 2x_{10}(6) + x_{12}(6),$$

$$x_5(3) = x_3(6) + x_6(6) + 2x_7(6) + x_9(6) + x_{10}(6) + x_{11}(6) + x_{12}(6),$$

$$x_6(3) = x_6(6) + x_{10}(6) + 2x_{11}(6) + 3x_{12}(6) + x_{13}(6).$$

The concentrations of the hexads are connected by two constraints:

$$x_1(6) + 6x_2(6) + 6x_3(6) + 6x_4(6) + 3x_5(6) + 6x_6(6) + 12x_7(6)$$
$$+ 2x_8(6) + 3x_9(6) + 6x_{10}(6) + 6x_{11}(6) + 6x_{12}(6) + x_{13}(6) = 1,$$

$$x_1(6) + 5x_2(6) + 4x_3(6) + 4x_4(6) + 2x_5(6) + 3x_6(6) + 6x_7(6) + x_8(6)$$
$$+ x_9(6) + 2x_{10}(6) + 2x_{11}(6) + x_{12}(6) = x.$$

Thus, there are 11 independent concentrations of the hexads, and the system of equations minimizing the free energy of mixing is given by:

$$\frac{\partial f^M}{\partial x_i} = 0, \quad (i = 1, \ldots, 11).$$

If composition is $x = \frac{1}{2}$, then the number of the independent concentrations of the hexads reduces, since $x_1(6) = x_{13}(6)$, $x_2(6) = x_{12}(6)$, $x_3(6) = x_{11}(6)$, $x_4(6) = x_{10}(6)$, and $x_5(6) = x_9(6)$. As the result, there are six independent variables that can be $x_1(6)$, $x_2(6)$, $x_3(6)$, $x_4(6)$, $x_5(6)$, and $x_6(6)$. Further evaluations reduce the number of the independent variables up to two variables. If $x_2(6)$ and $x_4(6)$ are the independent variables, the other clusters are:

$$x_1(6) = x_{13}(6) = \frac{1}{2} - 15x_2(6) - 15x_4(6) - \frac{x_4(6)^2}{x_2(6)},$$

$$x_3(6) = x_6(6) = x_{11}(6) = x_{12}(6) = x_2(6),$$

$$x_5(6) = x_7(6) = x_9(6) = x_{10}(6) = x_4(6),$$

$$x_8(6) = \frac{x_4(6)^2}{x_2(6)},$$

$$x_1(3) = x_6(3) = \frac{1}{2} = -9x_2(6) - 14x_4(6) - \frac{x_4(6)^2}{x_2(6)},$$

$$x_2(3) = x_5(3) = 4x_2(6) + 4x_4(6),$$

$$x_3(3) = x_4(3) = x_2(6) + 6x_4(6) + \frac{x_4(6)^2}{x_2(6)},$$

$$x_1(2) = x_3(2) = \frac{1}{2} - 5x_2(6) - 10x_4(6) - \frac{x_4(6)^2}{x_2(6)},$$

$$x_2(2) = 5x_2(6) + 10x_4(6) + \frac{x_4(6)^2}{x_2(6)}.$$

4.9.2 Miscibility Gap

The Helmholtz free energy of mixing of $A_x B_{1-x}$ regular solution is:

$$f^M = \alpha_{A-B} x_2(2) + RT \left[2 \sum_{i=1}^{13} \alpha_i(6) x_i(6) \ln x_i(6) - 6 \sum_{j=1}^{6} \alpha_j(3) x_j(3) \ln x_j(3) \right.$$

$$+ 2 \sum_{k=1}^{3} \alpha_k(2) x_k(2) \ln x_k(2) + 3x \ln x$$

$$\left. + 3(1-x) \ln(1-x) \right].$$

The miscibility gap is determined by minimizing the free energy of mixing of the two-phase system. The free energy of mixing is a symmetric function of the composition with respect to the point $x = \frac{1}{2}$, i.e., $f^M(x) = f^M(1-x)$. Accordingly, the miscibility gap boundary due to the constraint for the concentrations of the hexads given by:

$$\sum_{i=1}^{13} x_i(6) = 1 \text{ or } x_{13}(6) = 1 - \sum_{i=1}^{12} x_i(6)$$

is described by the system of equations:

$$\frac{df^M}{dx} = 0, \quad \frac{\partial f^M}{\partial x_i(6)} = 0, \quad (i = 1, ..., 12).$$

References

[1] R.J. Baxter, Exactly Solved Models in Statistical Mechanics, Academic Press, London, 1982, 1984, 1989.
[2] R. Kikuchi, A theory of cooperative phenomena, Phys. Rev. 81 (6) (1951) 988–1003.
[3] A.G. Schlijper, Convergence of the cluster-variation method in the thermodynamic limit, Phys. Rev. B 27 (11) (1983) 6841–6848.
[4] R. Kikuchi, CVM entropy algebra, Progr. Theoret. Phys. Suppl. 115 (1994) 1–26.
[5] K. Huang, Statistical Mechanics, John Wiley & Sons, New York, 1963, 1987 (Chapter 16).
[6] J.A. Baker, Methods of approximation in the theory of regular mixtures, Proc. Roy. Soc. A 216 (1953) 45–56.
[7] J.W. Gibbs, Collected Works, vol. 1, Yale University Press, New Haven, Connecticut, 1948.
[8] J.W. Cahn, On spinodal decomposition, Acta Metall. 9 (9) (1961) 795–801.
[9] J.W. Cahn, On spinodal decomposition in cubic crystals, Acta Metall. 10 (3) (1962) 179–183.

[10] G.B. Stringfellow, Spinodal decomposition and clustering in III/V alloys, J. Electron. Mater. 11 (5) (1982) 903–918.

[11] G.A. Korn, T.M. Korn, Mathematical Handbook, second ed., McGraw-Hill Book Company, New York, 1968 (Chapter 13).

[12] G.H. Wannier, Antiferromagnetism. The triangular ising net, Phys. Rev. 79 (2) (1950) 357–364.

[13] L. Onzager, Crystal statistics. I. A two-dimensional model with an order-disorder transition, Phys. Rev. 65 (3–4) (1944) 117–149.

5

Submolecular Regular Solutions

Alloys with different types of atoms in both sublattices play a very important role in solid-state electronics and, especially, in optoelectronics. However, they cannot be sufficiently reasonably represented as conventional regular solutions as described in Chapters 3 and 4 of this book. Chapter 5 is devoted to the modified regular solution model and consideration of the quaternary alloys with cations and anions of two types. The existence of two mixed sublattices stipulates the peculiarities absent in the alloys described up to now in this book. One of the most important peculiarities is the dependence of the properties of the alloys on the thermodynamic characteristics of the constituent compounds. The crystal structures of the alloys are, in fact, the mixtures of bonds corresponding to the constituent compounds. Therefore, these alloys are presented as submolecular alloys. More multicomponent alloys can be considered in a similar way to quaternary alloys described in this chapter.

The replacement of the values of the Gibbs free energy by the values of the Helmholtz free energy, and vice versa, established in Chapter 2, Section 2.9, is widely used in this chapter.

5.1 QUATERNARY REGULAR SOLUTIONS OF FOUR BINARY COMPOUNDS

$A_x^{II}B_{1-x}^{II}C_y^{VI}D_{1-y}^{VI}$, $A_x^{III}B_{1-x}^{III}C_y^{V}D_{1-y}^{V}$, and $A_x^{IV}B_{1-x}^{IV}C_y^{VI}D_{1-y}^{VI}$ semiconductor alloys can be considered as $A_xB_{1-x}C_yD_{1-y}$ quaternary regular solutions. Regular solutions of this class differ significantly from the $A_xB_yC_{1-x-y}D$ and $AB_xC_yD_{1-x-y}$ molecular quaternary regular solutions that, as shown in Chapter 3, Section 3.3.2, can be represented as quasi-ternary regular solutions. The crystal lattice of $A_xB_{1-x}C_yD_{1-y}$ regular solutions consists of two sublattices as well as the crystal lattice of $A_xB_yC_{1-x-y}D$ and $AB_xC_yD_{1-x-y}$ regular solutions. However, both sublattices of $A_xB_{1-x}C_yD_{1-y}$ regular solutions are mixed, since each sublattice fills with two types of atoms. Accordingly, any atom can have different-type atoms at the nearest surroundings. Let the cation sublattice fill with A, B atoms, and C, D atoms form the anion sublattice. The arrangement of cations and anions on the anion and cation lattice sites, respectively, are not considered. The electroneutrality condition demands the same numbers of cations and anions, i.e., $N_A + N_B = N_C + N_D = N$.

There are four kinds of pairs of the nearest neighbors (chemical bonds in $A_x^{II} B_{1-x}^{II} C_y^{VI} D_{1-y}^{VI}$, $A_x^{III} B_{1-x}^{III} C_y^V D_{1-y}^V$, and $A_x^{IV} B_{1-x}^{IV} C_y^{VI} D_{1-y}^{VI}$ semiconductor alloys) AC, AD, BC, and BD in such solutions. The numbers of the pairs are connected by only three equations:

$$N_{AC} + N_{AD} = z_1 N_A, \qquad (5.1.1a)$$

$$N_{AC} + N_{BC} = z_1 N_C, \qquad (5.1.1b)$$

$$N_{AC} + N_{AD} + N_{BC} + N_{BD} = z_1 N, \qquad (5.1.1c)$$

where z_1 is the first coordination number. Hence, $A_x B_{1-x} C_y D_{1-y}$ regular solutions are quaternary solutions of four binary compounds, AC, AD, BC, and BD, since they include the pairs of the nearest neighbors corresponding to four types of chemical bonds in $A_x^{II} B_{1-x}^{II} C_y^{VI} D_{1-y}^{VI}$, $A_x^{III} B_{1-x}^{III} C_y^V D_{1-y}^V$, and $A_x^{IV} B_{1-x}^{IV} C_y^{VI} D_{1-y}^{VI}$. Four kinds of pairs and only three equations describing their numbers lead to an absence of a one-to-one correspondence between the numbers of atoms and numbers of pairs of the nearest neighbors (bonds). In other words, the numbers of pairs are the functions of the numbers of atoms and numbers of certain types of pairs given by using Eqn (5.1.1) as (for example, the functions of the numbers of atoms and number of AC pairs):

$$N_{AD} = z_1 N_A - N_{AC}, \qquad (5.1.2a)$$

$$N_{BC} = z_1 N_C - N_{AC}, \qquad (5.1.2b)$$

$$N_{BD} = z_1 (N - N_A - N_C) + N_{AC}. \qquad (5.1.2c)$$

There are three equations similar to Eqn (5.1.1) for the atomic concentration (x and y) and concentration of the pairs (x_{AC}, x_{AD}, x_{BC}, and x_{BD}), which are written as:

$$x_{AC} + x_{AD} = x, \qquad (5.1.3a)$$

$$x_{AC} + x_{BC} = y. \qquad (5.1.3b)$$

$$x_{AC} + x_{AD} + x_{BC} + x_{BD} = 1 \qquad (5.1.3c)$$

and the concentrations of the pairs are expressed by using Eqn (5.1.3) as:

$$x_{AD} = x - x_{AC}, \qquad (5.1.4a)$$

$$x_{BC} = y - x_{AC}, \qquad (5.1.4b)$$

$$x_{BD} = 1 - x - y + x_{AC}. \qquad (5.1.4c)$$

This important peculiarity leads to the absence of a one-to-one correspondence between the atomic concentrations (x and y) and the concentrations of bonds (x_{AC}, x_{AD}, x_{BC}, and x_{BD}). The concentrations of atoms and pairs represent the elemental and chemical compositions, respectively, in the alloys described here. Due to the absence of the one-to-one correspondence, one elemental composition (x and y) corresponds to a vast set of the pair compositions (x_{AC}, x_{AD}, x_{BC}, and x_{BD}) or chemical compositions. It can be illustrated as follows. The exchange of lattice sites between the different type cations or anions can lead to the reaction between the bonds:

$$nAC + nBD \rightarrow nAD + nBC, \ n = 1, ..., z_1$$

or vice versa. The number n in reaction $nAC + nBD \rightarrow nAD + nBC$ depends on the nearest surroundings of cations or anions participating in the exchange. For regular solutions with the zinc blende and wurtzite structures, these reactions in the case of the exchange of cations A and B situated in the centers of the tetrahedral cells are given (as shown in Chapter 1, Section 1.9) by:

$$n = 1:$$

$$1A4C + 1B3C1D \rightarrow 1A3C1D + 1B4C,$$

$$1A3C1D + 1B2C2D \rightarrow 1A2C2D + 1B3C1D,$$

$$1A2C2D + 1B1C3D \rightarrow 1A1C3D + 1B2C2D,$$

$$1A1C3D + 1B4D \rightarrow 1A4D + 1B1C3D,$$

$$\text{reaction: } AC + BD \rightarrow AD + BC;$$

$$n = 2:$$

$$1A4C + 1B2C2D \rightarrow 1A2C2D + 1B4C,$$

$$1A3C1D + 1B1C3D \rightarrow 1A1C3D + 1B3C1D,$$

$$1A2C2D + 1B4D \rightarrow 1A4D + 1B2C2D,$$

$$\text{reaction: } 2AC + 2BD \rightarrow 2AD + 2BC;$$

$$n = 3:$$

$$1A4C + 1B1C3D \rightarrow 1A1C3D + 1B4C,$$

$$1A3C1D + 1A4D \rightarrow 1A4D + 1B3C1D,$$

$$\text{reaction: } 3AC + 3BD \rightarrow 3AD + 3BC;$$

$$n = 4:$$

$$1A4C + 1B4D \rightarrow 1A4D + 1B4C,$$

$$\text{reaction: } 4AC + 4BD \rightarrow 4AD + 4BC,$$

where $1A4C$ is the tetrahedral cell with central atom A. Reaction $nAC + nBD \rightarrow nAD + nBC$ changes the numbers of bonds in $A_xB_{1-x}C_yD_{1-y}$ regular solution, and thus it demonstrates that one elemental composition (x and y) corresponds to the very large quantity of the bond (pair) compositions (x_{AC}, x_{AD}, x_{BC} and x_{BD}). The reactions for the exchange of the lattice sites of anions C and D can be considered in a similar way. In fact, in the case of the exchanges of anions, cations A and B should be substituted for anions C and D, respectively; and anions C and D have to be replaced by cations A and B, correspondingly.

For $A_xB_{1-x}C_yD_{1-y}$ regular solutions with the rock salt structure, the reactions for the exchange of cations A and B situated in the centers of the octahedral cells are:

$$n = 1:$$

$$1A6C + 1B5C1D \rightarrow 1A5C1D + 1B6C,$$

$$1A5C1D + 1B4C2D \rightarrow 1A4C2D + 1B5C1D,$$

$$1A4C2D + 1B3C3D \rightarrow 1A3C3D + 1B4C2D,$$

$$1A3C3D + 1B2C4D \rightarrow 1A2C4D + 1B3C3D,$$

$$1A2C4D + 1B3C3D \rightarrow 1A3C3D + 1B2C4D,$$

$$1A1C5D + 1B6D \rightarrow 1A6D + 1B1C5D,$$

$$\text{reaction: } AC + BD \rightarrow AD + BC;$$

$$n = 2:$$

$$1A6C + 1B4C2D \rightarrow 1A4C2D + 1B6C,$$

$$1A5C1D + 1B3C3D \rightarrow 1A3C3D + 1B5C1D,$$

$$1A4C2D + 1B2C4D \rightarrow 1A2C4D + 1B4C2D,$$

$$1A3C3D + 1B1C5D \rightarrow 1A1C5D + 1B3C3D,$$

$$1A2C4D + 1B6D \rightarrow 1A6D + 1B4C2D,$$

$$\text{reaction: } 2AC + 2BD \rightarrow 2AD + 2BC;$$

$$n = 3:$$

$$1A6C + 1B3C3D \rightarrow 1A3C3D + 1B6C,$$

$$1A5C1D + 1A2C4D \rightarrow 1A2C4D + 1B5C1D,$$

$$1A4C2D + 1B1C5D \rightarrow 1A1C5D + 1B4C2D,$$

$$1A3C3D + 1B6D \rightarrow 1A6C + 1B3C3D,$$

$$\text{reaction: } 3AC + 3BD \rightarrow 3AD + 3BC;$$

$$n = 4:$$

$$1A6C + 1B2C2D \rightarrow 1A2C2D + 1B6C,$$

$$1A5C1D + 1B1C5D \rightarrow 1A1C5D + 1B5C1D,$$

$$1A4C2D + 1B6D \rightarrow 1A6D + 1B4C2D;$$

$$\text{reaction: } 4AC + 4BD \rightarrow 4AD + 4BC;$$

$$n = 5:$$

$$1A6C + 1B1C5D \rightarrow 1A1C5D + 1B6C,$$

$$1A5C1D + 1B6D \rightarrow 1A6D + 1B5C1D;$$

$$\text{reaction: } 5AC + 5BD \rightarrow 5AD + 5BC;$$

$$n = 6:$$

$$1A6C + 1B6D \rightarrow 1A6D + 1B6C,$$

$$\text{reaction: } 6AC + 6BD \rightarrow 6AD + 6BC,$$

where $1A6C$ is the octahedral cell with the central atom A. The reactions when exchanging the lattice sites of anions C and D can be also considered in a similar manner. Cations A and B should be replaced by anions C and D, respectively, and vice versa.

The model of $A_xB_{1-x}C_yD_{1-y}$ regular solutions differs significantly from the model of regular solutions described in Chapter 3. The model of $A_xB_{1-x}C_yD_{1-y}$ regular solutions is based on the following assumptions.

1. Cations and anions are distributed on the lattice sites of the cation and anion sublattices, respectively. Each cation and each anion may occupy any site on the cation and anion sublattices, correspondingly.
2. The lattice parameter of $A_xB_{1-x}C_yD_{1-y}$ regular solution does not depend on the distribution of atoms on the lattice sites and elemental composition (x,y). The internal degrees of freedom and degrees of freedom relating to the positions of atoms are separable. The partition function of $A_xB_{1-x}C_yD_{1-y}$ regular solution represented as a canonical ensemble is given by $Q = Q_{\text{Int.}}Q_{\text{Conf.}}Q_{\text{Ac}}$, where $Q_{\text{Int.}}$, $Q_{\text{Conf.}}$, and Q_{Ac} are, accordingly, the internal, configurational, and acoustic partition functions. Only the configurational and acoustic partition functions depend on the distribution of atoms on the lattice sites.
3. The internal energy corresponding to the configurational partition function depends on the interactions between the nearest and next-nearest atoms.

4. The interaction energy between the next-nearest neighbors depends on the type of atom situated between them. The interaction energies between the nearest and next-nearest atoms are independent on the surroundings.

5. The vibration motion of atoms is determined by pairs of the nearest neighbors.

The interaction energies between the next-nearest atoms are determined by the interaction parameters between binary compounds AC, AD, BC, and BD:

$$w_{AC-BC} = u_{ACB} - \frac{u_{ACA} + u_{BCB}}{2}, w_{AD-BD} = u_{ADB} - \frac{u_{ADA} + u_{BDB}}{2},$$

$$w_{AC-AD} = u_{CAD} - \frac{u_{CAC} + u_{DAD}}{2} \text{ and } w_{BC-BD} = u_{CBD} - \frac{u_{CBC} + u_{DBD}}{2},$$

where u_{ACB} is the interaction energy between atoms A and B with atom C between them. These interaction parameters are equivalent to the interaction parameters between the binary compounds in the molecular regular solutions of binary compounds considered in Chapter 3, Section 3.3.

The partition function of $A_xB_{1-x}C_yD_{1-y}$ regular solution represented as a canonical ensemble containing $N = N_A + N_B = N_C + N_D$ cations and anions is written as:

$$Q = Q_{\text{Int.}} Q_{\text{Conf.}} Q_{Ac}$$

$$= \prod_{i,j=1}^{2} \sigma_i^{N_i} \sigma_j^{N_j} \sum_{N_{ij}} \sum_{N_{iji}} \sum_{N_{jij}} g \exp\left\{-\frac{u_{ij}}{k_B T}\right\}^{N_{ij}}$$

$$\times \exp\left\{-\frac{u_{iji}}{k_B T}\right\}^{N_{iji}} \exp\left\{-\frac{u_{jij}}{k_B T}\right\}^{N_{jij}}$$

$$\times \prod_{l \neq i, m \neq j}^{2} \exp\left\{-\frac{u_{ijl}}{k_B T}\right\}^{\frac{N_{ijl}}{2}} \exp\left\{-\frac{u_{jim}}{k_B T}\right\}^{\frac{N_{jim}}{2} N_{ij}} q_{ij}^{N_{ij}}, \quad (5.1.5)$$

where i and l are the indexes of cations; j and m are the indexes of anions; σ_i and σ_j are the partition functions of the internal degrees of freedom of cations and anions, respectively; N_i and N_j are the numbers of the i-th type cations and j-th type anions, respectively; $g = g(N, N_i, N_j, N_{ij}, N_{iji}, N_{jij})$ is the number of the geometrically different configurations having the same value of the internal energy determined by the pairs and triads of atoms; u_{ij} is the interaction energy between the nearest atoms in compound ij; u_{ijl} is the interaction energy between the nearest cations of the i-th and l-th types with the j-th type anion between the cations; u_{jim} is the interaction energy between the nearest anions of the j-th and m-th types with cations of the i-th type between the anions; N_{ijl} is the number of i-j-l-th type cation–anion–cation triads in the regular solution; N_{jim} is the number of j-i-m-th type anion–cation–anion triads in the solution; and q_{ij} is the partition function of the vibration motion per the ij-th bond.

The Helmholtz free energy of $A_xB_{1-x}C_yD_{1-y}$ regular solution obtained from the partition function (5.1.5) is:

$$F = -k_BT \ln Q$$

$$= \sum_{i,j=1,l \neq i,m \neq j}^{2} \left\{ - N_ik_BT \ln \sigma_i - N_jk_BT \ln \sigma_j - k_BT \ln g + u_{ij}N_{ij} \right.$$

$$\left. + u_{iji}N_{iji} + u_{jij}N_{jij} + \frac{u_{ijl}N_{ijl} + u_{jim}N_{jim}}{2} - k_BT \ln q_{ij}^{N_{ij}} \right\}.$$

$$(5.1.6)$$

The Helmholtz free energy (Eqn (5.1.2)) can be rewritten as:

$$F = \sum_{i,j=1,l \neq i,m \neq j}^{2} \left[\mu_{ij}^0 \frac{N_{ij}}{z_1} + \frac{z_2}{2z_1}(w_{ij-lj}N_{ijl} + w_{ij-im}N_{jim}) - k_BT \ln g \right], \quad (5.1.7)$$

where $\mu_{ij}^0 = u_i + u_j - T(s_i + s_j) + z_1u_{ij} + z_2\frac{u_{iji} + u_{jij}}{2} - z_1k_BT \ln q_{ij}$ is the chemical potential per molecule of compound ij; u_i, u_j, s_i, and s_j are the energies and entropies corresponding to the internal degrees of freedom of the i-th type cations and j-th type anions, respectively; and $w_{ij-lj} = u_{ijl} - \frac{u_{iji} + u_{ljl}}{2}$ is the interaction parameter between binary compounds ij and lj in the regular solution. The Helmholtz free energy per mole is given by:

$$f = \sum_{i,j,l,m=1}^{2} \left[f_{ij}^0 x_{ij} + \frac{\alpha_{ij-lj}}{2} x_{ijl} + \frac{\alpha_{ij-im}}{2} x_{ijm} + RT \ln g \right],$$

$$f = \sum_{i,j,l,m=1}^{2} \left[\mu_{ij}^{0S} x_{ij} + \frac{\alpha_{ij-lj}}{2} x_{ijl} + \frac{\alpha_{ij-im}}{2} x_{ijm} + RT \ln g \right],$$

where μ_{ij}^0 is the molar chemical potential of compound ij; x_{ij} and x_{ijm} are the concentrations of pairs ij and triads ijm, respectively; and $\alpha_{ij-lj} = z_2N_{Av}w_{ij-lj}$ is the molar interaction parameter between compounds ij and lj.

5.2 MODIFIED BAKER'S APPROACH

The cluster variation method will be used to consider $A_xB_{1-x}C_yD_{1-y}$ submolecular regular solutions in this chapter. Baker's approach, which provides a simple way to obtain the configurational coefficients in the cluster variation method, is described in Chapter 4, Section 4.1. This approach allows us to consider elemental and molecular alloys as regular solutions. The consequence of this is that in Baker's approach, any atom of an elemental alloy and any atom of a mixed sublattice can be situated on any site of a lattice and any site of a mixed sublattice, respectively. The different situation occurs for $A_xB_{1-x}C_yD_{1-y}$ regular solutions. In this case, the crystal structure consists of two geometrically equivalent sublattices filled, correspondingly, with cations and anions. In other words, cations and anions can be situated on the sites of their sublattices. Thus, there are two kinds of one-point clusters—cations and anions. In the one-point approximation for $A_xB_{1-x}C_yD_{1-y}$ regular solutions, cations and anions mix freely, and in such a case, the number of configurations is:

$$g(1) = \frac{(N_A + N_B)!(N_C + N_D)!}{N_A!N_B!N_C!N_D!}. \tag{5.2.1}$$

Two-point clusters in such solutions are cation–anion pairs, and thus there is only one kind and four types of cation–anion pairs: AC, AD, BC, and BD. The two-point approximation for $A_xB_{1-x}C_yD_{1-y}$ regular solutions in the cluster variation method may be described as follows. The number of configurations in the two-point approximation $g(2)$ is a function of the number of pairs $N_p(2,1)$ ($p = 1,...,4$), number of cations $N_i(1,1)$, and number of anions $N_j(1,2)$, where the second number in the brackets represents the kind of cluster. The number of configurations $g(2)$ of $A_xB_{1-x}C_yD_{1-y}$ regular solution in the two-point approximation is thus:

$$g(2) = g(1)h(N_A, N_B, N_C, N_D)g(2,1) \tag{5.2.2}$$

where $h(N_A,N_B,N_C,N_D)$ is the normalizing factor that is required to give correctly the total number of configurations at a high temperature when cations and anions are distributed randomly, and $g(2,1)$ is the total number of configurations of randomly distributed pairs, which is written as:

$$g(2,1) = \frac{(N_{AC} + N_{AD} + N_{BC} + N_{BD})!}{N_{AC}!N_{AD}!N_{BC}!N_{BD}!}. \tag{5.2.3}$$

Thus, the quantity $g(2,1)$ should be multiplied by the normalizing factor, which depends only on $N_i(1,1)$ and $N_j(1,2)$. The normalizing factor $h(N_A,N_B,N_C,N_D)$ is given by:

$$h(N_A, N_B, N_C, N_D) = \frac{\left(z_1\frac{N_AN_C}{N}\right)!\left(z_1\frac{N_AN_D}{N}\right)!\left(z_1\frac{N_BN_C}{N}\right)!\left(z_1\frac{N_BN_D}{N}\right)!}{(z_1N)!}, \tag{5.2.4}$$

where $N = N_A + N_B = N_C + N_D$. Thus, the molar configurational entropy of $A_xB_{1-x}C_yD_{1-y}$ regular solution in the two-point approximation looks like this:

$$s = R \ln g = R\{(z_1 - 1)[x \ln x + (1 - x)\ln(1-x) + y \ln y + (1 - y)\ln(1 - y)] \\ - z_1(x_{AC} \ln x_{AC} + x_{AD} \ln x_{AD} + x_{BC} \ln x_{BC} + x_{BD} \ln x_{BD})\}$$

or

$$s = R \ln g(2) = R \left[\delta(2,1) \sum_{i=1}^{4} \alpha_i(2,1)x_i(2,1)\ln x_i(2,1) \right.$$

$$+ \delta(1,1) \sum_{j=1}^{2} \alpha_j(1,1)x_j(1,1)\ln x_j(1,1)$$

$$\left. + \delta(1,2) \sum_{k=1}^{2} \alpha_k(1,2)x_k(1,2)\ln x_k(1,2) \right],$$

where $\delta(2,1) = -z_1$, $\delta(1,1) = z_1 - 1$, and $\delta(1,2) = z_1 - 1$ are the entropy coefficients for pairs, cations, and anions, respectively; $\alpha_i(2,1)$ and $x_i(2,1)$ are the number of distinguishable configurations generated by the symmetry operations and concentration of the two-point cluster of the i-th type, respectively. The coefficients for the basic clusters in the two-point approximation for $A_xB_{1-x}C_yD_{1-y}$ regular solutions $\delta(2,1) = -z_1$ two times more than the coefficients for the basic clusters for A_xB_{1-x} binary regular solution in the two-point approximation, calculated by Baker's approach as $\delta(2,1) = -\frac{N(2,1)}{N_{Av}} = -\frac{z_1}{2}$. It follows from the number of atoms in 1 mol of $A_xB_{1-x}C_yD_{1-y}$ regular solution equal to $2N_{Av}$ where N_{Av} is the Avogadro number. For $A_xB_{1-x}C_yD_{1-y}$ regular solution, the coefficient of the basic clusters is calculated as $\delta(2,1) = -\frac{N(2,1)}{N_{Av}} = -z_1$. Thus, the formula deriving the molar configurational entropy of $A_xB_{1-x}C_yD_{1-y}$ regular solution by using Baker's approach should be modified for the calculation of the coefficients. This implies that it is necessary to use the Avogadro number but not the total number of atoms in 1 mol of the regular solution. Accordingly, the procedure for the calculation of the entropy coefficients of $A_xB_{1-x}C_yD_{1-y}$ regular solution is:

$$\delta(n, t) = -N(n, t)/N_{Av} \tag{5.2.5}$$

and:

$$\delta(r, t) = -\frac{N(r,t)}{N_{Av}} - \sum_{q=r+1}^{n} \sum_{s} M(r, t; q, s), \quad (1 \le r < n), \tag{5.2.6}$$

where $N(n,t)$ is the total number of the n-point basic clusters of t-th kind in 1 mol of the regular solution, $N(r,t)$ is the total number of the r-point subclusters of t-th kind in 1 mol of the regular solution, and $M(r,t;q,s)$ is the number of clusters (r,t) being in cluster (q,s), where kinds of clusters are represented by t and s. The molar configurational entropy is:

$$s = R \sum_t \sum_r \delta(r,t) \sum_i \alpha_i(r,t) x_i(r,t) \ln x_i(r,t).$$

5.3 ONE-POINT APPROXIMATION

5.3.1 Helmholtz Free Energy

Atoms are the basic clusters. There are two kinds of atoms. Cations are the first type of atoms and anions are atoms of the second kind. This approximation is equivalent to the strictly regular approximation in the theory of regular solutions. The types of atoms, concentrations, and numbers of distinguishable configurations of the clusters are shown in Tables 5.1 and 5.2.

The entropy coefficients for atoms $\delta(1,1)$ and $\delta(1,2)$ are given, respectively, by:

$$\delta(1,1) = -\frac{N(1,1)}{N_{Av}} = -1 \text{ and } \delta(1,2) = -\frac{N(1,2)}{N_{Av}} = -1.$$

Accordingly, the molar configurational entropy of $A_x B_{1-x} C_y D_{1-y}$ regular solution is

$$s = R \left[\delta(1,1) \sum_{i=1}^{2} \alpha_i(1,1) x_i(1,1) \ln x_i(1,1) \right.$$

$$\left. + \delta(1,2) \sum_{j=1}^{2} \alpha_j(1,2) x_j(1,2) \ln x(1,2) \right].$$

TABLE 5.1 The first type of one-point clusters

Number	Type	Concentration	Number of distinguishable configurations, $\alpha_1(1,1)$
1	A	$x_1(1,1)$	1
2	B	$x_2(1,1)$	1

TABLE 5.2 The second type of one-point clusters

Number	Type	Concentration	Number of distinguishable configurations, $\alpha_1(1,1)$
1	C	$x_1(1,2)$	1
2	D	$x_2(1,2)$	1

The entropy of mixing corresponds to the randomly arranged cations and anions.

The Helmholtz free energy of such solution is:

$$F = \sum_{i,j,l \neq i, m \neq j}^{2} \left[\mu_{ij}^0 \frac{N_{ij}}{z_1} + \frac{z_2}{2z_1} (w_{ij-lj} N_{ijl} + w_{ij-im} N_{jim}) - k_B T \ln g \right],$$

where i and l are the indexes of cations, j and m are the indexes of anions, and g is the number of the different configurations. The numbers of pairs and triads in the case of the random distribution of cations and anions are given, respectively, by:

$$N_{ij} = z_1 \frac{N_i N_j}{N}, \tag{5.3.1.1}$$

$$N_{iji} = z_2 \frac{N_i^2 N_j}{2N^2}, \tag{5.3.1.2}$$

$$N_{jij} = z_2 \frac{N_i N_j^2}{2N^2}, \tag{5.3.1.3}$$

$$N_{ijl} = z_2 \frac{N_i N_j N_l}{N^2}, \tag{5.3.1.4}$$

$$N_{jim} = z_2 \frac{N_i N_j N_m}{N^2}, \quad (j \neq m). \tag{5.3.1.5}$$

The Helmholtz free energy by using expressions (5.3.1.1–5.3.1.5) is represented as:

$$F = \sum_{i,j,l \neq i, m \neq j}^{2} \left[\left(\mu_{ij}^0 + z_2 \frac{w_{ij-lj} N_l + w_{ij-im} N_m}{2N} \right) \frac{N_i N_j}{N} \right.$$
$$\left. - k_B T (2N \ln N - N_i \ln N_i - N_j \ln N_j) \right],$$

where $\mu_{ij}^0 = u_i + u_j - T(s_i + s_j) + z_1 u_{ij} + z_2 \frac{u_{iji} + u_{jij}}{2}$ is the chemical potential per molecule of binary compound ij; and u_i and s_i are the energy and entropy of the internal degrees of freedom of i-th type cation, respectively. The molar Helmholtz free energy of $A_x B_{1-x} C_y D_{1-y}$ regular solution is:

$$f = \sum_{i,j,l,m=1}^{2} \left[\mu_{ij}^0 x_{ij} + \frac{\alpha_{ij-lj}}{2} x_{ijl} + \frac{\alpha_{ij-im}}{2} x_{ijm} + RT \ln g \right]$$

$$= \sum_{i,j,l,m=1}^{2} \left[\mu_{ij}^0 x_i x_j + \alpha_{ij-lj} \frac{x_i x_j x_l}{2} + \alpha_{ij-im} \frac{x_i x_j x_m}{2} + RT(x_i \ln x_i + x_j \ln x_j) \right],$$

where μ_{ij}^0 is the molar chemical potential of compound ij; x_{ij} and x_{ijm} are the concentrations of pairs ij and triads ijm, respectively; and $\alpha_{ij-lj} = z_2 N_{Av} w_{ij-lj}$ is the molar interaction parameter between compounds ij and lj. If all interaction parameters are equal to zero, such solutions are called "$A_x B_{1-x} C_y D_{1-y}$ ideal regular solutions."

The concentrations of bonds or the short-range order in accordance with the random distribution of cations and anions in their sublattices are given by:

$$x_{AC} = xy, \ x_{AD} = x(1-y), \ x_{BC} = (1-x)y \text{ and } x_{BD} = (1-x)(1-y).$$

Thus, this approximation provides a one-to-one correspondence between the atomic (elemental) and chemical (bond) compositions of $A_x B_{1-x} C_y D_{1-y}$ regular solution.

5.3.2 Miscibility Gap

The miscibility gap of $A_x B_{1-x} C_y D_{1-y}$ regular solutions exists due to the positive interaction parameters between the binary compounds that cause the tendency to phase separation. The miscibility gap will be described by using the minimization of the Helmholtz free energy of the heterogeneous system, which consists of two quaternary regular solutions corresponding to the decomposed semiconductor alloy and having the compositions x_1, y_1 and x_2, y_2, respectively, and average composition x, y. The partition function of the heterogeneous system consisting of $A_{x_1} B_{1-x_1} C_{y_1} D_{1-y_1}$ and $A_{x_2} B_{1-x_2} C_{y_2} D_{1-y_2}$ regular solutions and represented as a canonical ensemble in which the total numbers of the different types of cations and anions are given values is:

$$Q = Q_{Int}.Q_{Conf}.Q_{Ac}$$

$$= \prod_{n=1}^{2} \sigma_i^{N_{ni}} \sigma_j^{N_{nj}} \sum_{N_{ij}} \sum_{N_{iji}} \sum_{N_{jij}} g \exp\left\{ -\frac{u_{ij}}{k_B T} \right\}^{N_{nij}}$$

$$\times \exp\left\{ -\frac{u_{ijl}}{k_B T} \right\}^{N_{niji}} \exp\left\{ -\frac{u_{jij}}{k_B T} \right\}^{N_{njij}}$$

$$\times \prod_{l \neq i, m \neq j} \exp\left\{ -\frac{u_{iji}}{k_B T} \right\}^{\frac{N_{ijl}}{2}} \exp\left\{ -\frac{u_{jim}}{k_B T} \right\}^{\frac{N_{jim}}{2}} q_{ij}^{N_{ij}}, \qquad (5.3.2.1)$$

where N_n, N_{ni}, N_{nl}, N_{nj}, and N_{nm} are the total number of cations (anions), cations of the i-th and l-th types and anions of the j-th and m-th types in the n-th ($n = 1, 2$) phase, respectively. The partition function (5.3.2.1) by using the expressions (5.3.1.1–5.3.1.5) can be rewritten as:

$$Q = \prod_{n=1}^{2} \prod_{i,j=1,l,m=1}^{2} \frac{(N_n!)^2}{N_{ni}!N_{nj}!} \left[\exp\left\{ -\frac{u_{ij}}{k_B T} \right\} \right]^{z_1 \frac{N_{ni}N_{nj}}{N_n}}$$

$$\times \left[\exp\left\{ -\frac{N_{nl}u_{ijl} + N_{nm}u_{jim}}{k_B T} \right\} \right]^{z_2 \frac{N_{ni}N_{nj}}{2N_n^2}} q_{ij}^{z_1 \frac{N_{ni}N_{nj}}{N_n}}$$

$$\times \exp\left\{ -\frac{u_{iji}}{k_B T} \right\}^{z_2 \frac{N_{ni}^2 N_{nj}}{2N_n^2}} \exp\left\{ -\frac{u_{jij}}{k_B T} \right\}^{z_2 \frac{N_{ni}N_{nj}^2}{2N_n^2}}$$

$$\times \prod_{l \neq i, m \neq j} \left[\exp\left\{ -\frac{N_{nl}u_{ijl} + N_{nm}u_{jim}}{k_B T} \right\} \right]^{z_2 \frac{N_{ni}N_{nj}}{2N_n^2}} q_{ij}^{z_1 \frac{N_{ni}N_{nj}}{N_n}},$$

where N_n, N_{ni}, N_{nl}, N_{nj}, and N_{nm} are the total number of cations (anions), cations of the i-th and l-th types and anions of the j-th and m-th types in the n-th ($n = 1, 2$) phase, respectively. The Helmholtz free energy of the two-phase system is:

$$F = -k_B T \ln Q$$

$$= \sum_{n=1}^{2} \sum_{i,j,l,m=1}^{2} \left[\left(\mu_{ij}^0 + z_2 \frac{w_{ij-lj}N_{nl} + w_{ij-im}N_{nm}}{2N} \right) \frac{N_{ni}N_{nj}}{N} \right.$$

$$\left. - k_B T(2N \ln N - N_{ni} \ln N_{ni} - N_{nj} \ln N_{nj}) - k_B T \ln q_{ij}^{z_1 \frac{N_{ni}N_{nj}}{N_n}} \right].$$

There are three constraints between the numbers of cations (anions), cations A, and anions C, which look like this:

$$\psi_1 = N_1 + N_2 - N = 0,$$

$$\psi_2 = N_{1A} + N_{2A} - N_A = 0,$$

$$\psi_3 = N_{1C} + N_{2C} - N_C = 0.$$

The numbers N_1, N_2, N_{1A}, N_{2A}, N_{1C}, and N_{2C} are thus the reciprocally dependent variables. Therefore, the Lagrange method of the undetermined multipliers is used in order to minimize the Helmholtz free energy. The Lagrange function of this system is expressed by:

$$L = F + \lambda_1 \psi_1 + \lambda_2 \psi_2 + \lambda_3 \psi_3,$$

where λ_1 is the Lagrange undetermined multiplier. The minimization is represented by the following system of equations:

$$\frac{\partial L}{\partial N_1} = 0, \quad \frac{\partial L}{\partial N_2} = 0, \quad \frac{\partial L}{\partial N_{1A}} = 0, \quad \frac{\partial L}{\partial N_{2A}} = 0,$$

$$\frac{\partial L}{\partial N_{1C}} = 0, \quad \frac{\partial L}{\partial N_{2C}} = 0,$$

$$\psi_1 = 0, \quad \psi_2 = 0 \text{ and } \psi_3 = 0.$$

Finally, the miscibility gap is described by these equations:

$$\frac{\partial F}{\partial N_1} - \frac{\partial F}{\partial N_2} = 0, \quad \frac{\partial F}{\partial N_{1A}} - \frac{\partial F}{\partial N_{2A}} = 0, \quad \frac{\partial F}{\partial N_{1C}} - \frac{\partial F}{\partial N_{2C}} = 0. \qquad (5.3.2.2)$$

If the interaction parameters are equal to zero, the system of Eqns (5.3.2.2) has the form:

$$-\left(\mu_{AC}^0 - \mu_{AD}^0 - \mu_{BC}^0 + \mu_{BD}^0\right)(x_1 y_1 - x_2 y_2) + RT \ln \frac{(1-x_1)(1-y_1)}{(1-x_2)(1-y_2)} = 0,$$

$$(5.3.2.3a)$$

$$\left(\mu_{AC}^0 - \mu_{AD}^0 - \mu_{BC}^0 + \mu_{BD}^0\right)(y_1 - y_2) + RT \ln \frac{(1-x_1)x_2}{x_1(1-x_2)} = 0, \qquad (5.3.2.3b)$$

$$\left(\mu_{AC}^0 - \mu_{AD}^0 - \mu_{BC}^0 + \mu_{BD}^0\right)(x_1 - x_2) + RT \ln \frac{(1-y_1)y_2}{y_1(1-y_2)} = 0, \qquad (5.3.2.3c)$$

where $x_2 = \frac{x - \gamma x_1}{1-\gamma}$, $y_2 = \frac{y - \gamma y_1}{1-\gamma}$, $\gamma = \frac{N_1}{N}$, and x_1, y_1, and $\gamma = \frac{N_1}{N}$ are the independent variables. This system of equations has the solution $0 < x_1$, x_2, y_1, $y_2 < 1$ if $\mu_{AC}^0 - \mu_{AD}^0 - \mu_{BC}^0 + \mu_{BD}^0 \neq 0$. Thus, the miscibility gap of $A_x B_{1-x} C_y D_{1-y}$ regular solution should exist if $\mu_{AC}^0 - \mu_{AD}^0 - \mu_{BC}^0 + \mu_{BD}^0 \neq 0$ even if all interaction parameters are equal to zero. This case fundamentally differs from, for example, that of $A_x B_y C_{1-x-y} D$ quaternary molecular regular solution of three binary compounds AD, BD, and BC. In fact, for $A_x B_{1-x} C_y D_{1-y}$ regular solution, there is an additional cause of decomposition. It occurs due to the reaction between chemical bonds after the exchange of the lattice sites between cations or anions:

$$nAC + nBD \rightarrow nAD + nBC, \, n = 1,...,z_1.$$

The quantity $\mu_{AC}^0 - \mu_{AD}^0 - \mu_{BC}^0 + \mu_{BD}^0$ is the important characteristic of $A_x B_{1-x} C_y D_{1-y}$ regular solution.

In the general case, when the interaction parameters between the compounds are not equal to zero, there are the following additional expressions for the right parts of the equations of the system describing the miscibility gap:

$$\sum_{n=1}^{2}(-1)^{n+1}\{(\alpha_{AC-BC} + \alpha_{AC-AD})x_n y_n + (\alpha_{AD-BD}x_n + \alpha_{BC-BD}y_n)$$

$$\times (2 - x_n - y_n) - 2[\alpha_{AC-BC}y_n + \alpha_{AD-BD}(1 - y_n)]x_n(1 - x_n)$$

$$- 2[\alpha_{AD-AD}x_n + \alpha_{BC-BD}(1 - x_n)]y_n(1 - y_n)\},$$

$$\sum_{n=1}^{2}(-1)^{n+1}\{[\alpha_{AC-BC}y_n + \alpha_{AD-BD}(1 - y_n)](1 - 2x_n)$$

$$+ (\alpha_{AC-AD} - \alpha_{BC-BD})y_n(1 - y_n)\},$$

$$\sum_{n=1}^{2}(-1)^{n+1}\{[\alpha_{AC-AD}x_n + \alpha_{BC-BD}(1 - x_n)](1 - 2y_n)$$

$$+ (\alpha_{AC-BC} - \alpha_{AD-BD})x_n(1 - x_n)\}.$$

5.3.3 Quantity $\mu_{AC}^0 - \mu_{AD}^0 - \mu_{BC}^0 + \mu_{BD}^0$

The quantity $\mu_{AC}^0 - \mu_{AD}^0 - \mu_{BC}^0 + \mu_{BD}^0$ is the relation between the Gibbs free energies of the compounds. The standard enthalpies of formation (standard enthalpy change of formation or standard heat of formation) and heat capacities are available for a lot of semiconductor compounds. That is why the method of the estimation of the quantity $\mu_{AC}^0 - \mu_{AD}^0 - \mu_{BC}^0 + \mu_{BD}^0$ based on these characteristics is the most widely used and will be considered here. This method can be represented as follows.

The enthalpy of ij-th binary compound at absolute temperature T is:

$$h_{ij}(T) = h_{ij}(T = 298.15 \text{ K}) + \int_{298.15}^{T} c_{ij}^P(T)dT,$$

where $h_{ij}(T = 298.15 \text{ K})$ is the standard molar enthalpy of ij-th binary compound, and c_{ij}^P is the specific heat capacity at the constant pressure. The value of quantity $h_{AC} - h_{AD} - h_{BC} + h_{BD}$ is obtained as:

$$h_{AC} - h_{AD} - h_{BC} + h_{BD} = \sum_{i,j=1}^{2}(-1)^i(-1)^j\left[h_{ij}(298.15 \text{ K}) + \int_{298.15}^{T} c_{ij}^P dT\right]$$

$\sum_{i,j=1}^{2}(-1)^i(-1)^j h_{ij}(298.15 \text{ K})$ quantity is equal to $\sum_{i,j=1}^{2}(-1)^i(-1)^j h_{ij}^{0f}$ (298.15 K), where h_{ij}^{0f} is the standard enthalpy of formation of the ij-th compound. The standard enthalpy of formation of the ij-th compound is the difference between the enthalpy of the ij-th compound and the sum of the enthalpies of the i-th and j-th constituent elements at a temperature of 298.15 K (25 °C) and at 1 bar of pressure (STP). Accordingly, the values of the quantity $h_{AC} - h_{AD} - h_{BC} + h_{BD}$ can be estimated by the formula:

$$h_{AC} - h_{AD} - h_{BC} + h_{BD} = \sum_{i,j=1}^{2} (-1)^i (-1)^j \left[h_{ij}^{0f}(298.15 \text{ K}) + \int_{298.15}^{T} c_{ij}^p dT \right]$$

In order to estimate the quantity $\mu_{AC}^0 - \mu_{AD}^0 - \mu_{BC}^0 + \mu_{BD}^0$, the entropies of the constituent compounds should be taken into account. The standard entropies are also available for a lot of semiconductor compounds. The entropy of a binary compound at temperature T is written as:

$$s_{ij}(T) = s_{ij}(T = 298.15 \text{ K}) + \int_{298.15}^{T} \frac{c_{ij}^P(T)}{T} dT,$$

where $s_{ij}(T = 298.15 \text{ K})$ is the standard entropy of the ij-th binary compound. Accordingly, the values of the quantity $\mu_{AC}^0 - \mu_{AD}^0 - \mu_{BC}^0 + \mu_{BD}^0$ can be estimated by the formula:

$$\mu_{AC}^0 - \mu_{AD}^0 - \mu_{BC}^0 + \mu_{BD}^0 = \sum_{i,j=1}^{2} (-1)^i (-1)^j \left[h_{ij}^{0f}(298.15 \text{ K}) - T s_{ij}(298.15 \text{ K}) \right.$$

$$\left. + \int_{298.15}^{T} c_{ij}^p dT - T \int_{298.15}^{T} \frac{c_{ij}^p}{T} dT \right]$$

The standard enthalpies of formation play a main role in the last formula. Therefore, in many cases only they can be taken into account.

5.3.4 Spinodal Decomposition Range

The spinodal decomposition range of $A_x B_{1-x} C_y D_{1-y}$ regular solution is described as follows. Let us consider a process of an exchange of the atoms between regions 1 and 2:

$$A(1) \leftrightarrow B(2), \; C(1) \leftrightarrow D(2)$$

or simultaneous exchange of A and C atoms from region 1 with B and D atoms from region 2. Let $x - \delta x$, $y - \delta y$ is the composition in region 1 having volume γV after exchange of atoms A and C with atoms B and D between regions 1 and 2, where V is the volume of the initial homogeneous solution. The composition of region 2 after the exchange is equal to

$x + \frac{\gamma}{1-\gamma}\delta x$, $y + \frac{\gamma}{1-\gamma}\delta y$, where $\gamma = \frac{1}{2}$ as was established in Chapter 4, Section 4.2.3. We should derive the condition when this continuous change of the composition provides an increase of the Helmholtz free energy. The Helmholtz free energy of the system before the exchange of atoms is $f(x,y)$. The Helmholtz free energy of the system after the exchange is written as:

$$\frac{1}{2}[f(x+\delta x, y+\delta y) + f(x-\delta x, y-\delta y)].$$

The difference between the Helmholtz free energies of the system represented by the Tailor's multidimensional series expansion is:

$$\delta f = f(x,y) - \frac{1}{2}f(x+\delta x, y+\delta y) - \frac{1}{2}f(x-\delta x, y-\delta y) \approx \frac{1}{2}\frac{\partial^2 f(x,y)}{\partial x^2}\delta x^2$$

$$+ \frac{\partial^2 f(x,y)}{\partial x \partial y}\delta x \delta y + \frac{1}{2}\frac{\partial^2 f(x,y)}{\partial y^2}\delta y^2.$$

Thus, the difference between the Helmholtz free energies is expressed by the quadratic form. The quadratic form is the positive definite quadratic form if the following conditions are fulfilled:

$$\frac{\partial^2 f(x,y)}{\partial x^2} > 0, \qquad \begin{vmatrix} \dfrac{\partial^2 f(x,y)}{\partial x^2} & \dfrac{\partial^2 f(x,y)}{\partial x \partial y} \\[2mm] \dfrac{\partial^2 f(x,y)}{\partial x \partial y} & \dfrac{\partial^2 f(x,y)}{\partial y^2} \end{vmatrix} > 0.$$

The spinodal decomposition range is derived using the condition $\delta f = 0$. The quadratic form ceases to be a positive definite when one of the following two quantities becomes equal to zero:

$$\frac{\partial^2 f(x,y)}{\partial x^2}, \frac{\partial^2 f(x,y)}{\partial x^2}\frac{\partial^2 f(x,y)}{\partial y^2} - \left(\frac{\partial^2 f(x,y)}{\partial x \partial y}\right)^2, \qquad (5.3.4.1)$$

where

$$\frac{\partial^2 f(x,y)}{\partial x^2} = -2\alpha_{AC-BC}y - 2\alpha_{AD-BD}(1-y) + \frac{RT}{x(1-x)},$$

$$\frac{\partial^2 f(x,y)}{\partial y^2} = -2\alpha_{AC-AD}x - 2\alpha_{BC-BD}(1-x) + \frac{RT}{y(1-y)},$$

$$\frac{\partial^2 f(x,y)}{\partial x \partial y} = \mu_{AC}^0 - \mu_{AD}^0 - \mu_{BC}^0 + \mu_{BD}^0 + (\alpha_{AC-BC} - \alpha_{AD-BD})(1-2x)$$

$$+ (\alpha_{AC-AD} - \alpha_{BC-BD})(1-2y).$$

If the interaction parameters between the compounds are equal to zero, the spinodal decomposition range is described by the equation:

$$\left(\frac{\mu^0_{AC} - \mu^0_{AD} - \mu^0_{BC} + \mu^0_{BD}}{RT}\right)^2 - \frac{1}{x(1-x)y(1-y)} = 0.$$

This equation has solution $0 < x,\ y < 1$ if $\mu^0_{AC} - \mu^0_{AD} - \mu^0_{BC} + \mu^0_{BD} \neq 0$. Accordingly, again as in the miscibility gap case, the spinodal decomposition range of $A_xB_{1-x}C_yD_{1-y}$ regular solution should exist if $\mu^0_{AC} - \mu^0_{AD} - \mu^0_{BC} + \mu^0_{BD} \neq 0$ even if the interaction parameters between the compounds are equal to zero. It is also the consequence of the reaction between chemical bonds after the exchange of the lattice sites between cations or anions:

$$nAC + nBD \rightarrow nAD + nBC,\ n = 1,...,z_1.$$

As it was described in Chapter 4, Section 4.2.3, spinodal decomposition leads to the coherency strain energy in semiconductor alloys because of the different lattice parameters of the components. Therefore, the new phases that occurred are in the strained state. The initial stage of spinodal decomposition results in self-diffusion transfers of atoms at the distances of the order of a lattice parameter. In crystalline semiconductor alloys, the initial stage of spinodal decomposition results in self-diffusion transfers of atoms on the distances of the order of a lattice parameter in the planes, ensuring the minimal coherency strain energy. Thus, two thin layers form in the homogeneous alloy. The compositions of these layers at the initial stage are considered as constant values due to their small thickness. As is shown in Appendix 1, in alloys with the cubic structure, the occurred layers should be oriented in the {001} planes. The coherency strain energy of such an alloy can be rewritten as:

$$u^C = \frac{1}{2} \sum_{i=1}^{2} v_i \frac{\left(C^i_{11} - C^i_{12}\right)\left(C^i_{11} + 2C^i_{12}\right)}{C^i_{11}} \left(\frac{a_i - a}{a}\right)^2, \qquad (5.3.4.2)$$

where v_i, C^i_{11}, a_i, and a are molar volume, the stiffness coefficient, and lattice parameters of the i-th domain and homogeneous alloy, respectively. At the initial stage of spinodal decomposition, the molar volume, the stiffness coefficient, and lattice parameter of the i-th layer are very close, respectively, to those of the homogeneous alloy. Therefore, in Eqn (5.3.4.2) all multipliers except the last one depend insignificantly on the composition. Accordingly, in the estimates, the molar volumes and stiffness coefficients of the layers can be considered as the same quantities of the homogeneous alloy and the coherency strain energy can be rewritten as:

$$u^C = v \frac{(C_{11} - C_{12})(C_{11} + 2C_{12})}{C_{11}} \left[\frac{a(x,y) - a}{a}\right]^2,$$

where $v = v_{AC}xy + v_{AD}x(1-y) + v_{BC}(1-x)y + v_{BD}(1-x)(1-y)$, $C_{11} = C_{11}^{AC}xy + C_{11}^{AD}x(1-y) + C_{11}^{BC}(1-x)y + C_{11}^{BD}(1-x)(1-y)$, and $a = a_{AC}xy + a_{AD}x(1-y) + a_{BC}(1-x)y + a_{BD}(1-x)(1-y)$ are the molar volume, stiffness coefficient, and lattice parameter of the alloy, respectively, and are considered as constants; and v_{AC}, C_{11}^{AC}, and a_{AC} are the molar volume, stiffness coefficient, and lattice parameter of compound AC, correspondingly. The spinodal decomposition range of $A_xB_{1-x}C_yD_{1-y}$ semiconductor alloy is derived using the condition $\delta(f + u^C) = 0$ that is fulfilled when one of the two expressions becomes equal to zero:

$$\frac{\partial^2(f + u^C)}{\partial x^2},$$

$$\frac{\partial^2(f + u^C)}{\partial x^2}\frac{\partial^2(f + u^C)}{\partial y^2} - \left[\frac{\partial^2(f + u^C)}{\partial x \partial y}\right]^2,$$

where $\frac{\partial^2 u^C}{\partial x^2} = 2v\frac{(C_{11}-C_{12})(C_{11}+2C_{12})}{C_{11}}\left(\frac{a_{AD}-a_{BD}+\Delta ay}{a}\right)^2$, $\frac{\partial^2 u^C}{\partial y^2} = 2v\frac{(C_{11}-C_{12})(C_{11}+2C_{12})}{C_{11}}$

$\left(\frac{a_{BC}-a_{BD}+\Delta ax}{a}\right)^2\frac{\partial^2 u^C}{\partial x \partial y} = 2v\frac{(C_{11}-C_{12})(C_{11}+2C_{12})}{C_{11}} \times \frac{(a_{AD}-a_{BD}+\Delta ay)(a_{BC}-a_{BD}+\Delta ax)}{a^2}$ and

$\Delta a = a_{AC} - a_{AD} - a_{BC} + a_{BD}$.

5.3.5 Addition of Cation—Anion Pairs

The first considered distinctive peculiarity of $A_xB_{1-x}C_yD_{1-y}$ submolecular regular solutions from other quaternary regular solutions of binary compounds lies in the fact that their chemical composition cannot be obtained from the elemental composition. The other distinctive peculiarity is an occurrence of the different types of chemical bonds with the addition of the same cation—anion pairs to such solutions. It can be represented as follows.

Let us consider the addition of nA and nC atoms to $A_xB_{1-x}C_yD_{1-y}$ regular solution in which $N_A + N_B = N_C + N_D = N$ and $N \gg n$. The quantities of chemical bonds before the addition of A and C atoms are:

$$N_{AC}^{Before} = z_1\frac{N_AN_C}{N} = zNxy,$$

$$N_{AD}^{Before} = z_1\frac{N_A(N-N_C)}{N} = z_1Nx(1-y),$$

$$N_{BC}^{Before} = z_1\frac{(N-N_A)N_C}{N} = z_1N(1-x)y,$$

$$N_{BD}^{Before} = z_1\frac{(N-N_A)(N-N_C)}{N} = z_1N(1-x)(1-y).$$

The quantities of chemical bonds after the addition of nA and nC atoms to the regular solution are written as:

$$N_{AC}^{After} = z_1 \frac{(N_A + n)(N_C + n)}{N + n},$$

$$N_{AD}^{After} = z_1 \frac{(N_A + n)(N - N_C)}{N + n},$$

$$N_{BC}^{After} = z_1 \frac{(N - N_A)(N_C + n)}{N + n},$$

$$N_{BD}^{After} = z_1 \frac{(N - N_A)(N - N_C)}{N + n}.$$

The differences between the numbers of bonds after and before the addition of the atoms are:

$$\Delta N_{AC} = N_{AC}^{After} - N_{AC}^{Before} = z_1 \frac{(N_A + n)(N_C + n)}{N + n} - z_1 \frac{N_A N_C}{N}$$

$$= z_1 \frac{N(N_A + n)(N_C + n) - N_A N_C(N + n)}{N(N + n)} \approx z_1 \frac{N(N_A + N_C) - N_A N_C}{N^2}$$

$$= z_1(x + y - xy)n,$$

$$\Delta N_{AD} = N_{AD}^{After} - N_{AD}^{Before} \approx z_1 \frac{N^2 - N(N_A + N_C) + N_A N_C}{N^2}$$

$$= z_1(1 - x - y + xy)n,$$

$$\Delta N_{BC} = N_{BC}^{After} - N_{BC}^{Before} \approx z_1 \frac{N^2 - N(N_A + N_C) + N_A N_C}{N^2}$$

$$= z_1(1 - x - y + xy)n,$$

$$\Delta N_{BD} = N_{BD}^{After} - N_{BD}^{Before} \approx -z_1 \frac{N^2 - N(N_A + N_C) + N_A N_C}{N^2}$$

$$= -z_1(1 - x - y + xy)n,$$

$$\Delta N_{AC} + \Delta N_{AD} + \Delta N_{BC} + \Delta N_{BD} = z_1 n.$$

If $N_A = N_C = \frac{1}{2}N$, then:

$$\Delta N_{AC} = \frac{3}{4}z_1 n \left(z_1 = 4, \ \Delta N_{AC} = 3n; z_1 = 6, \ \Delta N_{AC} = \frac{9}{2}n \right),$$

$$\Delta N_{AD} = \frac{1}{4}z_1 n \left(z_1 = 4, \ \Delta N_{AD} = n; z_1 = 6, \ \Delta N_{AD} = \frac{3}{2}n \right),$$

$$\Delta N_{BC} = \frac{1}{4}z_1 n \left(z_1 = 4, \ \Delta N_{BC} = n; z_1 = 6, \ \Delta N_{BC} = \frac{3}{2}n \right),$$

$$\Delta N_{BD} = -\frac{1}{4}z_1 n \left(z_1 = 4, \ \Delta N_{BD} = -n; z_1 = 6, \ \Delta N_{BD} = -\frac{3}{2}n \right).$$

5.4 TWO-POINT APPROXIMATION

5.4.1 Helmholtz Free Energy

The cation–anion pairs are the basic clusters in the two-point approximation. There is one kind of the pairs. Cations are the first kind of atoms and anions are atoms of the second kind. This approximation is equivalent to the quasi-chemical approximation in the theory of regular solutions. The types of the pairs, atoms, concentrations, and numbers of distinguishable configurations of the clusters are shown in Tables 5.3–5.5. The concentrations of the basic clusters are considered as variables. The concentrations of the subclusters are:

$$x_1(1,1) = x_1(2,1) + x_2(2,1), x_2(1,1) = x_3(2,1) + x_4(2,1),$$
$$x_1(1,2) = x_1(2,1) + x_3(2,1), x_2(1,2) = x_2(2,1) + x_4(2,1).$$

There is the constraint between the variables that is given by:

$$x_1(2,1) + x_2(2,1) + x_3(2,1) + x_4(2,1) = 1.$$

If the composition of $A_xB_{1-x}C_yD_{1-y}$ regular solution is a given value, there are two additional constraints:

$$x_1(2,1) + x_2(2,1) = x \text{ and } x_1(2,1) + x_3(2,1) = y.$$

TABLE 5.3 Two-point clusters (cation–anion pairs)

Number	Type	Concentration	Number of distinguishable configurations, $\alpha_1(1,1)$
1	AC	$x_1(2,1)$	1
2	AD	$x_2(2,1)$	1
3	BC	$x_3(2,1)$	1
4	BD	$x_4(2,1)$	1

TABLE 5.4 One-point clusters of the first kind (cations)

Number	Type	Concentration	Number of distinguishable configurations, $\alpha_1(1,1)$
1	A	$x_1(1,1)$	1
2	B	$x_2(1,1)$	1

TABLE 5.5   One-point clusters of the second kind (anions)

Number	Type	Concentration	Number of distinguishable configurations, $\alpha_2(1,2)$
1	C	$x_1(1,2)$	1
2	D	$x_2(1,2)$	1

The entropy coefficients for the basic clusters and subclusters are written, respectively, as:

$$\delta(2,1) = -\frac{N(2,1)}{N_{Av}} = -z_1,$$

$$\delta(1,1) = -\frac{N(1,1)}{N_{Av}} - M(2,1;1,1)\delta(2,1) = z_1 - 1,$$

$$\delta(1,2) = -\frac{N(1,2)}{N_{Av}} - M(2,1;1,2)\delta(2,1) = z_1 - 1.$$

Accordingly, the configurational entropy looks like this:

$$
\begin{aligned}
s = R\{ & -z_1[x_1(2,1)\ln x_1(2,1) + x_2(2,1)\ln x_2(2,1) \\
& + x_3(2,1)\ln x_3(2,1) + x_4(2,1)\ln x_4(2,1)] \\
& + (z_1 - 1)[x_1(1,1)\ln x_1(1,1) + x_2(1,1)\ln x_2(1,1) \\
& + x_1(1,2)\ln x_1(1,2) + x_2(1,2)\ln x_2(1,2)]\}.
\end{aligned}
$$

The internal energy of $A_xB_{1-x}C_yD_{1-y}$ regular solution depends on the interactions between the nearest and next-nearest atoms and the interaction energy between the next-nearest atoms depends on the type of atom situated between them. In the cluster variation method, it is supposed that basic clusters are independent and randomly distributed objects. However, this assumption should be modified when calculating the concentrations of triads (Section 5.4.3). As a result, the concentrations of triads are:

$$x_{ijl} = \frac{x_{ij}x_{lj}}{x_j}, \quad x_{jim} = \frac{x_{ij}x_{im}}{x_i}.$$

The molar Helmholtz free energy of $A_xB_{1-x}C_yD_{1-y}$ regular solution has the form:

$$
f = \sum_{\substack{i,j, \\ i,m=1}}^{2} \left\{
\begin{aligned}
& \mu_{ij}^0 x_{ij} + \alpha_{ij-im}\frac{x_{ij}x_{im}}{2x_i} + \alpha_{ij-lj}\frac{x_{ij}x_{lj}}{2x_j} \\
& -RT[(1-z_1)(x_i \ln x_i + x_j \ln x_j) - z_1 x_{ij} \ln x_{ij}]
\end{aligned}
\right\},
$$

where μ_{ij}^0 is the chemical potential per ij-type molecule and α_{ij-im} is the interaction parameter between binary compounds of the ij-th and im-th types.

5.4.2 Short-Range Order

There are four types of atoms and four types of chemical bonds in $A_xB_{1-x}C_yD_{1-y}$ regular solutions. The numbers of atoms and chemical bonds are connected by three equations:

$$N_{AC} + N_{AD} = z_1 N_A,$$

$$N_{AC} + N_{BC} = z_1 N_C,$$

$$N_{AC} + N_{AD} + N_{BC} + N_{BD} = z_1(N_A + N_B) = z_1(N_C + N_D).$$

Thus, a one-to-one correspondence between the numbers of atoms and numbers of chemical bonds is missing, because there are three relations between four numbers of atoms and four numbers of chemical bonds. The concentrations of chemical bonds in solutions are normally called the short-range order. The short-range order and chemical composition are the same values in $A_xB_{1-x}C_yD_{1-y}$ regular solution because the chemical composition is expressed by the concentrations of four types of chemical bonds between the nearest atoms, which correspond to AC, AD, BC, and BD binary compounds. In the two-point approximation, the chemical composition is determined by the minimization of the Helmholtz free energy of solution. At the given value of the atomic composition, the concentrations of pairs of the nearest atoms depend on the concentrations of atoms and concentration of one-type pairs, which are as follows:

$$x_{AD} = x_A - x_{AC}, \; x_{BC} = x_C - x_{AC} \text{ and } x_{BD} = 1 - x - y + x_{AC}.$$

Thus, the Helmholtz free energy of $A_xB_{1-x}C_yD_{1-y}$ regular solution depends on one independent variable, x_{AC}. The minimum of the Helmholtz free energy is given by:

$$\frac{df}{dx_{AC}} = 0.$$

The Helmholtz free energy has the form:

$$f = \sum_{\substack{i,j, \\ i,m=1}}^{2} \left\{ \begin{array}{l} \mu_{ij}^0 x_{ij} + \alpha_{ij-im} \dfrac{x_{ij} x_{im}}{2x_i} + \alpha_{ij-lj} \dfrac{x_{ij} x_{lj}}{2x_j} \\[2mm] -RT[(1 - z_1)(x_i \ln x_i + x_j \ln x_j) - z_1 x_{ij} \ln x_{ij}] \end{array} \right\}$$

and its minimum condition is written as:

$$\mu_{AC}^0 - \mu_{AD}^0 - \mu_{BC}^0 + \mu_{BD}^0 + \alpha_{AC-AD}\frac{x - 2x_{AC}}{x}$$

$$+ \alpha_{AC-BC}\frac{y - 2x_{AC}}{y} - \alpha_{BC-BD}\frac{1 - x - 2y + 2x_{AC}}{1 - x}$$

$$- \alpha_{AD-BD}\frac{1 - 2x - y + 2x_{AC}}{1 - y} + z_1 RT \ln\frac{x_{AC}(1 - x - y + x_{AC})}{(x - x_{AC})(y - x_{AC})} = 0.$$

The minimum condition demonstrates that the concentrations of the bonds of $A_xB_{1-x}C_yD_{1-y}$ regular solution depend on the quantity $\mu_{AC}^0 - \mu_{AD}^0 - \mu_{BC}^0 + \mu_{BD}^0$, interaction parameters, and temperature. It is considered that if atoms are distributed randomly, then the short-range order is absent.

5.4.3 Miscibility Gap

The tendency to spinodal decomposition due to the positive interaction parameters provides the formation of the miscibility gap in $A_xB_{1-x}C_yD_{1-y}$ regular solutions. The description of the miscibility gap will be carried out using the minimization of the Helmholtz free energy of the heterogeneous system consisting of two quaternary regular solutions (decomposed solution) with the compositions x_1, y_1 and x_2, y_2, respectively, and average composition x, y. The partition function of the heterogeneous system consisting of $A_{x_1}B_{1-x_1}C_{y_1}D_{1-y_1}$ and $A_{x_2}B_{1-x_2}C_{y_2}D_{1-y_2}$ regular solutions represented as a canonical ensemble similar to Eqn (5.3.2.1) is:

$$Q = Q_{\text{Int.}}Q_{\text{Conf.}}Q_{\text{Ac}}$$

$$= \prod_{n=1}^{2}\prod_{i,j=1}^{2}\sigma_i^{N_{ni}}\sigma_j^{N_{nj}}\sum_{N_{ij}}\sum_{N_{iji}}\sum_{N_{jij}}g\exp\left\{-\frac{u_{ij}}{k_BT}\right\}^{N_{nij}}$$

$$\times \exp\left\{-\frac{u_{iji}}{k_BT}\right\}^{N_{niji}}\exp\left\{-\frac{u_{jij}}{k_BT}\right\}^{N_{njij}}$$

$$\times \prod_{l\neq i,m\neq j}\exp\left\{-\frac{u_{ijl}}{k_BT}\right\}^{\frac{N_{nijl}}{2}}\exp\left\{-\frac{u_{jim}}{k_BT}\right\}^{\frac{N_{njim}}{2}}q_{ij}^{N_{nij}}, \qquad (5.4.3.1)$$

where N_n, N_{ni}, N_{nl}, N_{nj}, N_{nm}, N_{nij}, N_{nijl}, and N_{njim} are the total number of cations (anions), cations of the i-th and l-th types, anions of the j-th and m-th types, the ij-th type pairs, and the ijl-th and jim-th types of triads in the n-th ($n = 1, 2$) phase, respectively.

The number of the different configurations g in the partition function (5.4.3.1) in accordance with the formulas (5.2.1–5.2.4) is:

$$g(N, N_i, N_j, N_{ij}) = (N!)^2 \prod_{i,j=1}^{2} \frac{\left(z_1 \frac{N_i N_j}{N}\right)!}{N_i! N_j! N_{ij}!}. \qquad (5.4.3.2)$$

The number of the iji-th triads is obtained as follows. The number of the iji-th triads is written as:

$$N_{iji} = \frac{z_2 N}{2} P_{iji},$$

where P_{iji} is the probability of the disposition of the iji-th triad on the given lattice sites in the determined order defined as:

$$P_{iji} = \frac{N_{ij}}{z_1 N} \frac{N_{ij}}{z_1 N_j} = \frac{N_{ij}^2}{z_1^2 N N_j}.$$

Thus, the number of the iji-th triads is:

$$N_{iji} = \frac{z_2 N_{ij}^2}{2 z_1^2 N_j}. \qquad (5.4.3.3)$$

The number of the jij-th triads obtained in the same way is:

$$N_{jij} = \frac{z_2 N_{ij}^2}{2 z_1^2 N_i}. \qquad (5.4.3.4)$$

The number of the ijl-th ($l \neq i$) triads is obtained in a similar way. The ijl-th and lji-th ($l \neq i$) type triads situated on the same lattice sites are considered here as different. The number of such ijl-th type triads is:

$$N_{ijl}^* = \frac{z_2 N}{2} P_{ijl}^*,$$

where P_{ijl}^* is the probability of the disposition of the ijl-th triad on the definite lattice sites given by:

$$P_{ijl}^* = \frac{N_{ij}}{z_1 N} \frac{N_{lj}}{z_1 N_j} = \frac{N_{ij} N_{lj}}{z_1^2 N_j}.$$

Accordingly, the number of the ijl-th triads is expressed by:

$$N_{ijl}^* = \frac{z_2 N_{ij} N_{lj}}{2 z_1^2 N_j}. \qquad (5.4.3.5)$$

Finally, the total number, or the number of the ijl-th and lji-th ($l \neq i$) type triads on the same lattice sites, is:

$$N_{ijl} = \frac{z_2 N_{ij} N_{lj}}{z_1^2 N_j}. \qquad (5.4.3.6)$$

The total number of the jim-th ($j \neq m$) triads derived in the same manner is:

$$N_{jim} = \frac{z_2 N_{ij} N_{im}}{z_1^2 N_i}. \tag{5.4.3.7}$$

The partition function of the two-phase system (5.4.3.1) by using formulas (5.1.1, 5.4.3.2–5.4.3.7) can be rewritten as:

$$Q = \prod_{n=1}^{2} \prod_{\substack{i,j, \\ l,m=1}}^{2} \sigma_i^{N_{ni}} \sigma_j^{N_{nj}} \frac{(N_n!)^2 \left(z_1 \frac{N_{ni}N_{nj}}{N_n}\right)!}{N_{ni}! N_{nj}!}$$

$$\times \sum_{N_{nij}} \frac{\exp\left\{-\frac{u_{ij}}{k_B T}\right\}^{N_{nij}} \exp\left\{-\frac{N_{ni}N_{nlj}u_{ijl} + N_{nj}N_{nim}u_{jim}}{k_B T}\right\}^{\frac{z_2 N_{nij}}{2z_1^2 N_{ni}N_{nj}}}}{N_{nij}} q_{ij}^{N_{nij}}. \tag{5.4.3.8}$$

The immediate use of formula (5.4.3.8) as well as the partition function (3.1.3) in Chapter 3 is inconvenient, since the thermodynamic quantities should be also obtained from the natural logarithm of the partition function containing the sum. Therefore, as well as in Chapter 3, the maximal term in the sum in Eqn (5.4.3.7) will be taken into account instead of the sum. Accordingly, the partition function (5.4.3.8) can be rewritten as:

$$Q = \prod_{n=1}^{2} \prod_{\substack{i,j, \\ l,m=1}}^{2} \sigma_i^{N_{ni}} \sigma_j^{N_{nj}} \frac{(N_n!)^2 \left(z_1 \frac{N_{ni}N_{nj}}{N_n}\right)!}{N_{ni}! N_{nj}! N_{nij}^{\#}}$$

$$\times \exp\left\{-\frac{u_{ij}}{k_B T}\right\}^{N_{nij}^{\#}} \exp\left\{-\frac{N_{ni}N_{nlj}^{\#}u_{ijl} + N_{nj}N_{njm}^{\#}u_{jim}}{k_B T}\right\}^{\frac{z_2 N_{nij}^{\#}}{2z_1^2 N_{ni}N_{nj}}} q_{ij}^{N_{nij}^{\#}}, \tag{5.4.3.9}$$

where $N_{nij}^{\#}$ are the numbers of the ij-th pairs in the maximal terms in the sums in Eqn (5.4.3.8).

The Helmholtz free energy of the two-phase system is:

$$F = \sum_{n=1}^{2} \sum_{i,j,l,m=1}^{2} \Bigg[-k_B T N_{ni} \ln \sigma_{ni} - k_B T N_{nj} \ln \sigma_{nj} + N_{nij}^{\#} u_{nij}$$

$$+ \frac{z_2}{2z_1^2} \left(\frac{N_{nij}^{\#} N_{nlj}^{\#}}{N_{nj}} u_{ijl} + \frac{N_{nij}^{\#} N_{nim}^{\#}}{N_{ni}} u_{jim}\right) - k_B T N_{nij}^{\#} \ln q_{ij}$$

$$- k_B T \Bigg(2 N_n \ln N_n - N_{ni} \ln N_{ni} - N_{nj} \ln N_{nj}$$

$$+ z_1 \frac{N_{ni}N_{nj}}{N_n} \ln z_1 \frac{N_{ni}N_{nj}}{N_n} - N_{nij}^{\#} \ln N_{nij}^{\#} \Bigg) \Bigg], \tag{5.4.3.10}$$

that can be rewritten as:

$$F = \sum_{n=1}^{2} \sum_{i,j,l,m=1}^{2} \left[\frac{N_{nij}^{\#}}{z_1} \mu_{ij}^0 + \frac{z_2}{2z_1^2} \left(\frac{N_{nij}^{\#} N_{nlj}^{\#}}{N_{nj}} w_{ij-lj} + \frac{N_{nij}^{\#} N_{nim}^{\#}}{N_{ni}} w_{ij-im} \right) \right.$$

$$- k_B T \left(2N_n \ln N_n - N_{ni} \ln N_{ni} - N_{nj} \ln N_{nj} \right.$$

$$\left. \left. + z_1 \frac{N_{ni} N_{nj}}{N_n} \ln z_1 \frac{N_{ni} N_{nj}}{N_n} - N_{nij}^{\#} \ln N_{nij}^{\#} \right) \right].$$

(5.4.3.11)

where $\mu_{ij}^0 = u_i + u_j - T(s_i + s_j) + z_1 u_{ij} + z_2 \frac{u_{iji} + u_{ijl}}{2} - z_1 k_B T \ln q_{ij}$ is the chemical potential of the ij-th compound, and $w_{ij-lj} = u_{ijl} - \frac{u_{iji} + u_{ijl}}{2}$ is the interaction parameter between binary compounds ij and lj.

If N, N_A, and N_C are chosen as given values, then there are three constraints for the numbers of atoms in the two-phase system, written as:

$$\psi_1 = N_1 + N_2 - N = 0,$$

$$\psi_2 = N_{1A} + N_{2A} - N_A = 0,$$

$$\psi_3 = N_{1C} + N_{2C} - N_C = 0.$$

The numbers of atoms N_1, N_2, N_{1A}, N_{2A}, N_{1C}, and N_{2C} are thus the reciprocally dependent variables. Moreover, the numbers of AC pairs in both phases are also the variables, since they are determined by the minimization of the free energy. The Lagrange method of the undetermined multipliers is used in order to minimize the Helmholtz free energy. The Lagrange function of the heterogeneous system is expressed by:

$$L - F + \lambda_1 \psi_1 + \lambda_2 \psi_2 + \lambda_3 \psi_3,$$

where λ_1 is the Lagrange undetermined multiplayer. The minimization is represented by the following system of equations:

$$\frac{\partial L}{\partial N_1} = 0, \quad \frac{\partial L}{\partial N_2} = 0, \quad \frac{\partial L}{\partial N_{1A}} = 0, \quad \frac{\partial L}{\partial N_{2A}} = 0,$$

$$\frac{\partial L}{\partial N_{1C}} = 0, \quad \frac{\partial L}{\partial N_{2C}} = 0,$$

$$\psi_1 = 0, \quad \psi_2 = 0 \text{ and } \psi_3 = 0,$$

$$\frac{\partial F}{\partial N_{1AC}^{\#}} = 0, \quad \frac{\partial F}{\partial N_{2AC}^{\#}} = 0.$$

Finally, the miscibility gap is described by the system of five equations:

$$\frac{\partial F}{\partial N_1} - \frac{\partial F}{\partial N_2} = 0, \qquad (5.4.3.12a)$$

$$\frac{\partial F}{\partial N_{1A}} - \frac{\partial F}{\partial N_{2A}} = 0, \tag{5.4.3.12b}$$

$$\frac{\partial F}{\partial N_{1C}} - \frac{\partial F}{\partial N_{2C}} = 0, \tag{5.4.3.12c}$$

$$\frac{\partial F}{\partial N_{1AC}^{\#}} = 0, \tag{5.4.3.12d}$$

$$\frac{\partial F}{\partial N_{2AC}^{\#}} = 0. \tag{5.4.3.12e}$$

If the interaction parameters between the compounds are equal to zero, the system of Eqns (5.4.3.12a–5.4.3.12e) is:

$$\left[\frac{(1-x_1)(1-y_1)}{(1-x_2)(1-y_2)}\right]^{z_1-1} - \left(\frac{1-x_1-y_1+x_{1AC}^{\#}}{1-x_2-y_2+x_{2AC}^{\#}}\right)^{z_1} = 0, \tag{5.4.3.13a}$$

$$\left[\frac{(1-x_1)x_2}{x_1(1-x_2)}\right]^{z_1-1} - \left[\frac{\left(1-x_1-y_1+x_{1AC}^{\#}\right)\left(x_1-x_{1AC}^{\#}\right)}{\left(1-x_2-y_2+x_{2AC}^{\#}\right)\left(x_2-x_{2AC}^{\#}\right)}\right]^{z_1} = 0, \tag{5.4.3.13b}$$

$$\left[\frac{(1-y_1)y_2}{y_1(1-y_2)}\right]^{z_1-1} - \left[\frac{\left(1-x_1-y_1+x_{1AC}^{\#}\right)\left(y_1-x_{1AC}^{\#}\right)}{\left(1-x_2-y_2+x_{2AC}^{\#}\right)\left(y_2-x_{2AC}^{\#}\right)}\right]^{z_1} = 0, \tag{5.4.3.13c}$$

$$\mu_{AC}^0 - \mu_{AD}^0 - \mu_{BC}^0 + \mu_{BD}^0 + z_1 RT \ln\frac{x_{1AC}(1-x_1-y_1+x_{1AC})}{(x_1-x_{1AC})(y_1-x_{1AC})}, \tag{5.4.3.13d}$$

$$\mu_{AC}^0 - \mu_{AD}^0 - \mu_{BC}^0 + \mu_{BD}^0 + z_1 RT \ln\frac{x_{2AC}(1-x_2-y_2+x_{2AC})}{(x_2-x_{2AC})(y_2-x_{2AC})}, \tag{5.4.3.13e}$$

where $x_2 = \frac{x-\gamma x_1}{1-\gamma}$, $y_2 = \frac{y-\gamma y_1}{1-\gamma}$, $\gamma = \frac{N_1}{N}$ and $x_1, y_1, \gamma = \frac{N_1}{N}$, $x_{1AC}^{\#}$, and $x_{2AC}^{\#}$ are the independent variables.

The system of Eqns (5.4.3.13a–5.4.3.13e) has the solution at $0 < x_1, x_2, y_1, y_2 < 1$ if $\mu_{AC}^0 - \mu_{AD}^0 - \mu_{BC}^0 + \mu_{BD}^0 \neq 0$ as well as the system of Eqns (5.3.2.3a–5.3.2.3c) in the description of the miscibility gap in the one-point approximation. Thus, the miscibility gap of $A_xB_{1-x}C_yD_{1-y}$ regular solutions should exist if $\mu_{AC}^0 - \mu_{AD}^0 - \mu_{BC}^0 + \mu_{BD}^0 \neq 0$ even if all interaction parameters between the compounds are equal to zero.

In the general case, when the interaction parameters are not equal to zero, the system of Eqns (5.4.3.12.a–5.4.3.12e) is:

$$\alpha_{BC-BD}\left[\left(\frac{y_1-x_{1AC}}{1-x_1}\right)^2 - \left(\frac{y_2-x_{2AC}}{1-x_2}\right)^2\right]$$

$$+ \alpha_{AD-BD}\left[\left(\frac{x_1-x_{1AC}}{1-y_1}\right) - \left(\frac{x_2-x_{2AC}}{1-y_2}\right)\right]$$

$$- RT\left[(z_1-1)\ln\frac{(1-x_1)(1-y_1)}{(1-x_2)(1-y_2)} - z_1 \ln\frac{1-x_1-y_1+x_{1AC}}{1-x_2-y_2+x_{2AC}}\right] = 0,$$

$$\alpha_{AC-AD}\left[\left(\frac{x_{1AC}}{x_1}\right)^2 - \left(\frac{x_{2AC}}{x_2}\right)^2\right] - \alpha_{BC-BD}\left[\left(\frac{y_1 - x_{1AC}}{1 - x_1}\right)^2 - \left(\frac{y_2 - x_{2AC}}{1 - x_2}\right)^2\right]$$

$$+ \alpha_{AD-BD}\left[\left(\frac{1 - 2x_1 - y_1 + 2x_{1AC}}{1 - y_1}\right)^2 - \left(\frac{1 - 2x_2 - y_2 + 2x_{2AC}}{1 - y_2}\right)^2\right]$$

$$+ RT\left[(z_1 - 1)\ln\frac{(1 - x_1)x_2}{x_1(1 - x_2)} - z_1\ln\frac{(1 - x_1 - y_1 + x_{1AC})(x_2 - x_{2AC})}{(1 - x_2 - y_2 + x_{2AC})(x_1 - x_{1AC})}\right]$$

$$= 0,$$

$$\alpha_{AC-BC}\left[\left(\frac{x_{1AC}}{y_1}\right)^2 - \left(\frac{x_{2AC}}{y_2}\right)^2\right] - \alpha_{AD-BD}\left[\left(\frac{x_1 - x_{1AC}}{1 - y_1}\right)^2 - \left(\frac{x_2 - x_{2AC}}{1 - y_2}\right)^2\right]$$

$$+ \alpha_{BC-BD}\left[\left(\frac{1 - x_1 - 2y_1 + 2x_{1AC}}{1 - x_1}\right)^2 - \left(\frac{1 - x_2 - 2y_2 + 2x_{2AC}}{1 - x_2}\right)^2\right]$$

$$+ RT\left[(z_1 - 1)\ln\frac{(1 - y_1)y_2}{y_1(1 - y_2)} - z_1\ln\frac{(1 - x_1 - y_1 + x_{1AC})(y_2 - x_{2AC})}{(1 - x_2 - y_2 + x_{2AC})(y_1 - x_{1AC})}\right]$$

$$= 0,$$

$$\mu^0_{AC} - \mu^0_{AD} - \mu^0_{BC} + \mu^0_{BD} + \alpha_{AC-AD}\frac{x_n - 2x_{nAC}}{x_n} + \alpha_{AC-BC}\frac{y_n - 2x_{nAC}}{y_n}$$

$$- \alpha_{BC-BD}\frac{1 - x_n - 2y_n + 2x_{nAC}}{1 - x_n} - \alpha_{AD-BD}\frac{1 - 2x_n - y_n + 2x_{nAC}}{1 - y_n}$$

$$+ z_1 RT\ln\frac{x_{nAC}(1 - x_n - y_n + x_{nAC})}{(x_n - x_{nAC})(y_n - x_{nAC})} = 0, \quad n = 1, 2,$$

where $x_2 = \frac{x - \gamma x_1}{1 - \gamma}$, $y_2 = \frac{y - \gamma y_1}{1 - \gamma}$, $\gamma = \frac{N_1}{N}$, and x_1, y_1, $\gamma = \frac{N_1}{N}$, x_{iAC}, and ($i = 1, 2$) are also the independent variables. Thus, the miscibility gap of $A_xB_{1-x}C_yD_{1-y}$ regular solution depends on the Helmholtz free energies of constituent compounds AC, AD, BC, and BD, interaction parameters between the compounds, and temperature.

5.4.4 Addition of Cation–Anion Pairs

Let us consider an addition of nA and nC atoms to $A_xB_{1-x}C_yD_{1-y}$ regular solution in which $N_A + N_B = N_C + N_D = N$ and $N \gg n$. To simplify the formulas, it is supposed that the interaction parameters between the compounds are equal to zero. The number N_{AC} of AC pairs in this case is obtained minimizing the Helmholtz free energy and is written as:

$$\beta = \frac{(z_1 N_A - N_{AC})(z_1 N_C - N_{AC})}{N_{AC}(z_1 N - z_1 N_A - z_1 N_C + N_{AC})}, \quad \text{where}$$

$$\beta = \exp\left\{ \frac{\mu_{AC}^0 - \mu_{AD}^0 - \mu_{BC}^0 + \mu_{BD}^0}{z_1 RT} \right\}.$$

Let be $\beta < 1$; then the numbers of pairs before the addition of nA and nC atoms are:

$$N_{AC}^{\text{Before}} = z_1 \frac{N_A + N_C + \beta(N - N_A - N_C)}{2(1 - \beta)}$$

$$- z_1 \sqrt{\left[\frac{N_A + N_C + \beta(N - N_A - N_C)}{2(1 - \beta)} \right]^2 - \frac{N_A N_C}{1 - \beta}}$$

$$N_{AD}^{\text{Before}} = z_1 N_A - z_1 \frac{N_A + N_C + \beta(N - N_A - N_C)}{2(1 - \beta)}$$

$$+ \sqrt{\left[z_1 \frac{N_A + N_C + \beta(N - N_A - N_C)}{2(1 - \beta)} \right]^2 - z_1^2 \frac{N_A N_C}{1 - \beta}},$$

$$N_{BC}^{\text{Before}} = z_1 N_C - z_1 \frac{N_A + N_C + \beta(N - N_A - N_C)}{2(1 - \beta)}$$

$$+ \sqrt{\left[z_1 \frac{N_A + N_C + \beta(N - N_A - N_C)}{2(1 - \beta)} \right]^2 - z_1^2 \frac{N_A N_C}{1 - \beta}},$$

$$N_{BD}^{\text{Before}} = z_1(N - N_A - N_C) + z_1 \frac{N_A + N_C + \beta(N - N_A - N_C)}{2(1 - \beta)}$$

$$- \sqrt{\left[z_1 \frac{N_A + N_C + \beta(N - N_A - N_C)}{2(1 - \beta)} \right]^2 - z_1^2 \frac{N_A N_C}{1 - \beta}}$$

The numbers of pairs after the addition of nA and nC atoms are written as:

$$N_{AC}^{\text{After}} = z_1 \frac{N_A + N_C + 2n + \beta(N - N_A - N_C - n)}{2(1 - \beta)}$$

$$- \sqrt{\left[z_1 \frac{N_A + N_C + 2n + \beta(N - N_A - N_C - n)}{2(1 - \beta)} \right]^2 - z_1^2 \frac{(N_A + n)(N_C + n)}{1 - \beta}},$$

$$N_{AD}^{\text{After}} = z_1(N_A + n) - z_1 \frac{N_A + N_C + 2n + \beta(N - N_A - N_C - n)}{2(1 - \beta)}$$

$$+ \sqrt{\left[z_1 \frac{N_A + N_C + 2n + \beta(N - N_A - N_C - n)}{2(1 - \beta)} \right]^2 - z_1^2 \frac{(N_A + n)(N_C + n)}{1 - \beta}},$$

$$N_{BC}^{After} = z_1(N_C + n) - z_1 \frac{N_A + N_C + 2n + \beta(N - N_A - N_C - n)}{2(1 - \beta)}$$

$$+ \sqrt{\left[z_1 \frac{N_A + N_C + 2n + \beta(N - N_A - N_C - n)}{2(1 - \beta)}\right]^2 - z_1^2 \frac{(N_A + n)(N_C + n)}{1 - \beta}},$$

$$N_{BD}^{After} = z_1(N - N_A - N_C - n) + z_1 \frac{N_A + N_C + 2n + \beta(N - N_A - N_C - n)}{2(1 - \beta)}$$

$$- \sqrt{[z_1 \frac{N_A + N_C + 2n + \beta(N - N_A - N' - n)}{2(1 - \beta)}]^2 - z_1^2 \frac{(N_A + n)(N_C + n)}{1 - \beta}}.$$

The differences between the numbers of pairs after and before the addition of atoms are

$$\Delta N_{AC} = N_{AC}^{After} - N_{AC}^{Before}$$

$$= z_1 n \left\{ \frac{2 - \beta}{2(1 - \beta)} - \frac{2\beta(1 - \beta)[(2 - \beta) - (1 - \beta)(x + y)]}{\sqrt{[\beta + (1 - \beta)(x + y)]^2 - 4(1 - \beta)xy}} \right\}, \quad (5.4.4.1)$$

$$\Delta N_{AD} = N_{AD}^{After} - N_{AD}^{Before}$$

$$= z_1 n \left\{ 1 - \frac{2 - \beta}{2(1 - \beta)} + \frac{2\beta(1 - \beta)[(2 - \beta) - (1 - \beta)(x + y)]}{\sqrt{[x + y + \beta(1 - x - y)]^2 - 4(1 - \beta)xy}} \right\},$$

$$(5.4.4.2)$$

$$\Delta N_{BC} = N_{BC}^{After} - N_{BC}^{Before}$$

$$= z_1 n \left\{ 1 - \frac{2 - \beta}{2(1 - \beta)} + \frac{2\beta(1 - \beta)[(2 - \beta) - (1 - \beta)(x + y)]}{\sqrt{[x + y + \beta(1 - x - y)]^2 - 4(1 - \beta)xy}} \right\},$$

$$(5.4.4.3)$$

$$\Delta N_{BD} = N_{BD}^{After} - N_{BD}^{Before}$$

$$= -z_1 n \left\{ 1 - \frac{2 - \beta}{2(1 - \beta)} + \frac{2\beta(1 - \beta)[(2 - \beta) - 2(1 - \beta)(x + y)]}{\sqrt{[x + y + \beta(1 - x - y)]^2 - 4(1 - \beta)xy}} \right\}.$$

$$(5.4.4.4)$$

The differences (5.4.4.1−5.4.4.4) depend on the composition of the regular solution, temperature, and chemical potentials of the constituent compounds. Evidently, in the general case, when the interaction parameters are not equal to zero, the differences between the numbers of pairs after and before the addition of atoms should depend also on the interaction parameters.

5.5 SIX-POINT APPROXIMATION

5.5.1 Helmholtz Free Energy

The $A_xB_{1-x}C_yD_{1-y}$ regular solutions with the zinc blende lattice are considered here. The hexad of atoms (six-point clusters) is chosen as a basic cluster. The hexad of atoms is a closed chain containing six pairs of the nearest neighbors. There are 20 types of hexads of the one kind in $A_xB_{1-x}C_yD_{1-y}$ regular solutions with the zinc blende lattice. The angles (three-point clusters or triads of atoms), pairs (two-point clusters), and atoms (one-point clusters) are subclusters, since they are the overlapping figures of the hexads. There are two kinds of angles. The first type of angle is the cation–anion–cation triad. The angle of the second type is the anion–cation–anion triad. There is one kind of pair, which is the cation–anion pair. Cations are atoms of the first kind and anions are the second kind of atoms. The types, concentrations, and number of configurations of the clusters are shown in Tables 5.6–5.11.

The calculation of the entropy coefficients of the six-, three-, two-, and one-point clusters by the modified Baker's approach are written, respectively, as:

$$\delta(6,1) = -\frac{N(6,1)}{N} = -4,$$

$$\delta(3,1) = -\frac{N(3,1)}{N} - M(3,1;6,1)\gamma(6,1) = 6,$$

$$\delta(3,2) = -\frac{N(3,2)}{N} - M(3,2;6,1)\gamma(6,1) = 6,$$

$$\delta(2,1) = -\frac{N(2,1)}{N} - M(2,1;6,1)\delta(6,1) - M(2,1;3,1)\delta(3,1)$$
$$- M(2,1;3,2)\delta(3,2) = -4,$$

$$\delta(1,1) = -\frac{N(1,1)}{N} - M(1,1;6,1)\delta(6,1) - M(1,1;3,1)\delta(3,1)$$
$$- M(1,1;3,2)\delta(3,2) - M(1,1;2,1)\delta(2,1) = -3,$$

$$\delta(1,2) = -\frac{N(1,2)}{N} - M(1,2;6,1)\delta(6,1) - M(1,2;3,1)\delta(3,1)$$
$$- M(1,2;3,2)\delta(3,2) - M(1,2;2,1)\delta(2,1) = -3.$$

TABLE 5.6 Six-point clusters

Number	Type	Probability	Number of different configurations, $\alpha_i(6,1)$
1	ACACAC	$x_1(6,1)$	1
2	ACACAD	$x_2(6,1)$	3
3	ACADAD	$x_3(6,1)$	3
4	ADADAD	$x_4(6,1)$	1
5	ACACBC	$x_5(6,1)$	3
6	ACACBD	$x_6(6,1)$	6
7	ACBCAD	$x_7(6,1)$	3
8	ACADBD	$x_8(6,1)$	3
9	ACBDAD	$x_9(6,1)$	6
10	ADADBD	$x_{10}(6,1)$	3
11	ACBCBC	$x_{11}(6,1)$	3
12	ACBCBD	$x_{12}(6,1)$	6
13	ACBDBC	$x_{13}(6,1)$	3
14	ACBDBD	$x_{14}(6,1)$	6
15	ADBCBD	$x_{15}(6,1)$	3
16	ADBDBD	$x_{16}(6,1)$	3
17	BCBCBC	$x_{17}(6,1)$	1
18	BCBCBD	$x_{18}(6,1)$	3
19	BCBDBD	$x_{19}(6,1)$	3
20	BDBDBD	$x_{20}(6,1)$	1

TABLE 5.7 The first type of three-point clusters

Number	Type	Probability	Number of different configurations, $\alpha_j(3,1)$
1	ACA	$x_1(3,1)$	1
2	ADA	$x_4(3,1)$	1
3	ACB	$x_2(3,1)$	2
4	ADB	$x_5(3,1)$	2
5	BCB	$x_3(3,1)$	1
6	BDB	$x_6(3,1)$	1

TABLE 5.8 The second type of three-point clusters

Number	Type	Probability	Number of different configurations, $\alpha_k(3,2)$
1	CAC	$x_1(3,2)$	1
2	CBC	$x_4(3,2)$	1
3	CAD	$x_2(3,2)$	2
4	CBD	$x_5(3,2)$	2
5	DAD	$x_3(3,2)$	1
6	DBD	$x_6(3,2)$	1

TABLE 5.9 Two-point clusters

Number	Type	Probability	Number of different configurations, $\alpha_l(2,1)$
1	AC	$x_1(2,1)$	1
2	AD	$x_2(2,1)$	1
3	BC	$x_3(2,1)$	1
4	BD	$x_4(2,1)$	1

TABLE 5.10 The first type of one-point clusters

Number	Type	Probability	Number of different configurations, $\alpha_m(1,1)$
1	A	$x_1(1,1)$	1
2	B	$x_2(1,1)$	1

TABLE 5.11 The second type of one-point clusters

Number	Type	Probability	Number of different configurations, $\alpha_n(1,2)$
1	C	$x_1(1,2)$	1
2	D	$x_2(1,2)$	1

The molar configurational entropy is given by:

$$s = R\left[-4\sum_{i=1}^{20}\alpha_i(6,1)x_i(6,1)\ln x_i(6,1) + \sum_{j=1}^{6}\alpha_j(3,1)x_j(3,1)\ln x_j(3,1)\right.$$

$$+ 6\sum_{k=1}^{6}\alpha_k(3,2)x_k(3,2)\ln x_k(3,2) - 4\sum_{l=1}^{4}\alpha_l(2,1)x_l(2,1)\ln x_l(2,1)$$

$$\left. -3\sum_{m=1}^{2}\alpha_m(1,1)x_m(1,1)\ln x_m(1,1) - 3\sum_{n=1}^{2}\alpha_n(1,2)x_n(1,2)\ln x_n(1,2)\right].$$

The molar internal energy is equal to the sum of the internal energies of pairs and triads:

$$u = \sum_{l=1}^{4}u_l(2,1)x_l(2,1) + \sum_{j=1}^{6}u_j(3,1)x_j(3,1) + \sum_{k=1}^{6}u_k(3,2)x_k(3,2)$$

$$= \sum_{l=1}^{4}u_l(2,1)x_l(2,1) + w_{AC-BC}x_3(3,1) + w_{AD-BD}x_4(3,1)$$

$$+ w_{AC-AD}x_3(3,2) + w_{BC-BD}x_4(3,2)$$

where $u_l(2,1)$ is the molar internal energy of the l-th cation–anion pair, $u_j(3,1)$ and $u_k(3,2)$ are the molar internal energies of the j-th cation–anion–cation and the k-th anion–cation–anion triads, respectively, and w_{AC-BC} is the interaction parameter between constituent compounds AC and BC.

The concentrations of the basic clusters are the variables. The concentrations of the subclusters are expressed by the concentrations of the basic clusters. The composition of the regular solution is:

$$x_1(1,1) = x_1(6,1) + 3x_2(6,1) + 3x_3(6,1) + x_4(6,1) + 2x_5(6,1) + 4x_6(6,1)$$

$$+ 2x_7(6,1) + 2x_8(6,1) + 4x_9(6,1) + 2x_{10}(6,1) + x_{11}(6.1)$$

$$+ 2x_{12}(6,1) + x_{13}(6,1) + 2x_{14}(6,1) + x_{15}(6,1) + x_{16}(6,1),$$

$$x_2(1,1) = 1 - x_1(1,1),$$

$$x_1(1,2) = x_1(6,1) + 2x_2(6,1) + x_3(6,1) + 3x_5(6,1) + 3x_6(6,1) + 2x_7(6,1)$$

$$+ x_8(6,1) + 2x_9(6,1) + 4x_{11}(6,1) + 4x_{12}(6,1) + 2x_{13}(6,1)$$

$$+ 2x_{14}(6,1) + x_{15}(6,1) + x_{17}(6,1) + 2x_{18}(6,1) + x_{19}(6,1),$$

$$x_2(1,2) = 1 - x_1(1,2).$$

The concentrations of the pairs have the form:

$$x_1(2,1) = x_1(6,1) + 2x_2(6,1) + x_3(6,1) + 2x_5(6,1) + 3x_6(6,1) + x_7(6,1)$$
$$+ x_8(6,1) + x_9(6,1) + x_{11}(6,1) + x_{12}(6,1) + x_{13}(6,1)$$
$$+ x_{14}(6,1),$$

$$x_2(2,1) = x_2(6,1) + 2x_3(6,1) + x_4(6,1) + x_6(6,1) + x_7(6,1) + x_8(6,1)$$
$$+ 3x_9(6,1) + 2x_{10}(6,1) + x_{12}(6,1) + +x_{14}(6,1) + x_{15}(6,1)$$
$$+ x_{16}(6,1),$$

$$x_3(2,1) = x_5(6,1) + x_6(6,1) + x_7(6,1) + x_9(6,1) + 2x_{11}(6,1) + 2x_{11}(6,1)$$
$$+ 3x_{12}(6,1) + x_{13}(6,1) + x_{14}(6,1) + x_{15}(6,1) + +x_{17}(6,1)$$
$$+ 2x_{18}(6,1) + x_{19}(6,1),$$

$$x_4(2,1) = x_6(6,1) + x_8(6,1) + x_9(6,1) + x_{10}(6,1) + x_{12}(6,1) + x_{13}(6,1)$$
$$+ 3x_{14}(6,1) + x_{15}(6,1) + 2x_{16}(6,1) + +x_{18}(6,1) + 2x_{19}(6,1)$$
$$+ x_{20}(6,1).$$

The concentrations of the triads are expressed by the concentrations of the hexads:

$$x_1(3,1) = x_1(6,1) + 2x_2(6,1) + x_3(6,1) + x_5(6,1) + 2x_6(6,1) + x_8(6,1),$$
$$x_2(3,1) = x_2(6,1) + 2x_3(6,1) + x_4(6,1) + x_7(6,1) + 2x_9(6,1) + x_{10}(6,1),$$
$$x_3(3,1) = x_5(6,1) + x_6(6,1) + x_7(6,1) + x_9(6,1) + x_{11}(6,1) + x_{12}(6,1)$$
$$+ x_{13}(6,1) + x_{14}(6,1),$$

$$x_4(3,1) = x_6(6,1) + x_8(6,1) + x_9(6,1) + x_{10}(6,1) + x_{12}(6,1) + x_{14}(6,1)$$
$$+ x_{15}(6,1) + x_{16}(6,1),$$

$$x_5(3,1) = x_{11}(6,1) + 2x_{12}(6,1) + x_{15}(6,1) + x_{17}(6,1) + 2x_{18}(6,1)$$
$$+ x_{19}(6,1),$$

$$x_6(3,1) = x_{13}(6,1) + 2x_{14}(6,1) + x_{16}(6,1) + x_{18}(6,1) + 2x_{19}(6,1)$$
$$+ x_{20}(6,1),$$

$$x_1(3,2) = x_1(6,1) + x_2(6,1) + 2x_5(6,1) + 2x_6(6,1) + x_{11}(6,1) + x_{13}(6,1),$$

$$x_2(3,2) = x_5(6,1) + x_7(6,1) + 2x_{11}(6,1) + 2x_{12}(6,1) + x_{17}(6,1) + x_{18}(6,1),$$

$$x_3(6,1) = x_2(6,1) + x_3(6,1) + x_6(6,1) + x_7(6,1) + x_9(6,1) + x_{11}(6,1)$$
$$+ x_{12}(6,1) + x_{14}(6,1),$$

$$x_4(3,2) = x_6(6,1) + x_9(6,1) + x_{12}(6,1) + x_{13}(6,1) + x_{14}(6,1) + x_{15}(6,1)$$
$$+ x_{18}(6,1) + x_{19}(6,1),$$

$$x_5(3,2) = x_3(6,1) + x_4(6,1) + 2x_9(6,1) + 2x_{10}(6,1) + x_{15}(6,1) + x_{16}(6,1),$$

$$x_6(3,2) = x_{13}(6,1) + 2x_{14}(6,1) + x_{16}(6,1) + x_{18}(6,1) + 2x_{19}(6,1)$$
$$+ x_{20}(6,1).$$

The concentrations of the basic clusters are connected by one constraint if the composition is not a given value. This constraint is:

$$\sum_{i=1}^{20} \alpha_i(6,1)x_i(6,1) = 1.$$

If the composition of $A_xB_{1-x}C_yD_{1-y}$ regular solution is a given value, there are two additional constraints for the probabilities of the basic clusters:

$$x_1(6,1) + 3x_2(6,1) + 3x_3(6,1) + x_4(6,1) + 2x_5(6,1) + 4x_6(6,1) + 2x_7(6,1)$$
$$+ 2x_8(6,1) + 4x_9(6,1) + 2x_{10}(6,1) + x_{11}(6,1) + 2x_{12}(6,1) + x_{13}(6,1)$$
$$+ 2x_{14}(6,1) + x_{15}(6,1) + x_{16}(6,1) = x$$

$$x_1(6,1) + 2x_2(6,1) + x_3(6,1) + 3x_5(6,1) + 3x_6(6,1) + 2x_7(6,1) + x_8(6,1)$$
$$+ 2x_9(6,1) + 4x_{11}(6,1) + 4x_{12}(6,1) + 2x_{13}(6,1) + 2x_{14}(6,1) + x_{15}(6,1)$$
$$+ x_{17}(6,1) + 2x_{18}(6,1) + x_{19}(6,1) = y.$$

The minimization of the Helmholtz free energy of $A_xB_{1-x}C_yD_{1-y}$ regular solution with the given composition can be done by the system of equations:

$$\frac{\partial f[x, y, x_i(6,1)]}{\partial x_i(6,1)} = 0, \quad (i = 1, \ldots, 17),$$

where $f = u - Ts$ and $x_i(6,1)$ are the independent variables.

5.5.2 Miscibility Gap

The miscibility gap of $A_xB_{1-x}C_yD_{1-y}$ regular solution is obtained by minimizing the Helmholtz free energy of two-phase system consisting of $A_{x_1}B_{1-x_1}C_{y_1}D_{1-y_1}$ and $A_{x_2}B_{1-x_2}C_{y_2}D_{1-y_2}$ regular solutions. There are three constraints for the numbers of atoms:

$$\psi_1 = N_1 + N_2 - N = 0,$$
$$\psi_2 = N_{1A} + N_{2A} - N_A = 0,$$
$$\psi_3 = N_{1C} + N_{2C} - N_C = 0,$$

where the total numbers of cations (anions), atoms A, and atoms C in the heterogeneous system represented by N, N_A, and N_C are given values; and N_1, N_{1A}, and N_{1C} are the numbers of cations (anions), atoms A, and atoms C in the first phase, respectively. N_1, N_2, N_{1A}, N_{2A}, N_{1C}, and N_{2C} are thus the reciprocally dependent variables. The concentrations of the basic clusters $x_{pi}(6,1)$ and subclusters $x_{pj}(3,1)$, $x_{pk}(3,2)$, $x_{pl}(2,1)$, $x_{pm}(1,1)$, and $x_{pn}(1,2)$ in the p-th ($p = 1$, 2) phase are given, respectively, by:

$$x_{pi}(6,1) = \frac{N_{pi}(6,1)}{z_1 N_p}, \quad x_{pj}(3,1) = \frac{2N_{pj}(3,1)}{z_2 N_p}, \quad x_{pk}(3,2) = \frac{2N_{pk}(3,2)}{z_2 N_p},$$

$$x_{pl}(2,1) = \frac{N_{pl}(2,1)}{z_1 N_p} x_{pm}(1,1) = \frac{N_{pm}(1,1)}{N_p}, \quad x_{pn}(1,2) = \frac{N_{pn}(1,1)}{N_p},$$

where N_p is the total number of cations (anions) in the p-th phase; and $N_{pi}(6,1)$, $N_{pj}(3,1)$, $N_{pk}(3,2)$, $N_{pl}(2,1)$, $N_{pm}(1,1)$, and $N_{pn}(1,2)$ are the numbers of the i-th hexads, the j-th and k-th triads, l-th pairs, and m-th and n-th atoms, respectively, in the p-th phase.

The Helmholtz free energy of the two-phase system is:

$$F = \sum_{p=1}^{2} \left[\sum_{l=1}^{4} u_l(2,1)N_{pl}(2,1) + \sum_{j=1}^{6} u_j(3,1)N_{pj}(3,1) + \sum_{k=1}^{6} u_k(3,2)N_{pk}(3,2) \right]$$

$$+ k_B T \sum_{p=1}^{2} \sum_{i=1}^{20} \alpha_i(6,1)N_{pi}(6,1)\ln\frac{N_{pi}(6,1)}{4N_p}$$

$$- k_B T \sum_{p=1}^{2} \sum_{j=1}^{6} \alpha_j(3,1)N_{pj}(3,1)\ln\frac{N_{pj}(3,1)}{6N_p}$$

$$- k_B T \sum_{p=1}^{2} \sum_{k=1}^{6} \alpha_k(3,2)N_{pk}(3,2)\ln\frac{N_{pk}(3,2)}{6N_p}$$

$$+ k_B T \sum_{p=1}^{2} \sum_{l=1}^{4} \alpha_l(2,1)N_{pl}(2,1)\ln\frac{N_{pl}(2,1)}{4N_p}$$

$$+ 3k_B T \sum_{p=1}^{2} \sum_{m=1}^{2} \alpha_m(1,1)N_{pm}(1,1)\ln\frac{N_{pm}(1,1)}{N_p}$$

$$+ 3k_B T \sum_{p=1}^{2} \sum_{n=1}^{2} \alpha_n(1,2)N_{pn}(1,2)\ln\frac{N_{pn}(1,2)}{N_p},$$

where u_l, u_j, and u_k are the internal energies of the l-th pair and j-th and k-th triads, respectively, $u_j = u_k = 0$ (j, $k = 1, 2, 5, 6$). There are six constraints for the numbers of the hexads in two phases:

$$\sum_{i=1}^{20} \alpha_i(6,1)N_{pi}(6,1) = 4N_p, \quad (p = 1,2),$$

$$
\begin{aligned}
&N_{p1}(6,1) + 3N_{p2}(6,1) + 3N_{p3}(6,1) + N_{p4}(6,1) + 2N_{p5}(6,1) + 4N_{p6}(6,1) \\
&\quad + 2N_{p7}(6,1) + 2N_{p8}(6,1) + 4N_{p9}(6,1) + 2N_{p10}(6,1) + N_{p11}(6,1) \\
&\quad + 2N_{p12}(6,1) + N_{p13}(6,1) + 2N_{p14}(6,1) + N_{p15}(6,1) + N_{p16}(6,1) = 4xN_p,
\end{aligned}
$$

$$
\begin{aligned}
&N_{p1}(6,1) + N_{p2}(6,1) + N_{p3}(6,1) + 3N_{p5}(6,1) + 3N_{p6}(6,1) + 2N_{p7}(6,1) \\
&\quad + N_{p8}(6,1) + 2N_{p9}(6,1) + 4N_{p11}(6,1) + 4N_{p12}(6,1) + 2N_{p13}(6,1) \\
&\quad + 2N_{p14}(6,1) + N_{p15}(6,1) + N_{p17}(6,1) + 2N_{p18}(6,1) + N_{p19}(6,1) = 4yN_p.
\end{aligned}
$$

The total numbers of cations (anions), atoms A, and atoms C, as well as the numbers of hexads in the phases, are the variables. The Lagrange method of the undetermined multipliers is used in order to obtain the minimization of the Helmholtz free energy. The Lagrange function of the heterogeneous system is:

$$L = F + \lambda_1\psi_1 + \lambda_2\psi_2 + \lambda_3\psi_3,$$

where λ_1 is the Lagrange undetermined multiplayer. The minimization is represented by the system of equations:

$$\frac{\partial L}{\partial N_1} = 0, \quad \frac{\partial L}{\partial N_2} = 0, \quad \frac{\partial L}{\partial N_{1A}} = 0, \quad \frac{\partial L}{\partial N_{2A}} = 0,$$

$$\frac{\partial L}{\partial N_{1C}} = 0, \quad \frac{\partial L}{\partial N_{2C}} = 0,$$

$$\psi_1 = 0, \quad \psi_2 = 0 \quad \text{and} \quad \psi_3 = 0,$$

$$\frac{\partial F}{\partial N_{pi}(6,1)} = 0, \quad (p = 1,2; i = 1, ..., 17).$$

Finally, the miscibility gap is described by the system of equations given by

$$\frac{\partial F}{\partial N_1} - \frac{\partial F}{\partial N_2} = 0, \quad \frac{\partial F}{\partial N_{1A}} - \frac{\partial F}{\partial N_{2A}} = 0, \quad \frac{\partial F}{\partial N_{1C}} - \frac{\partial F}{\partial N_{2C}} = 0,$$

$$\frac{\partial F}{\partial N_{pi}(6,1)} = 0, \quad (p = 1,2; i = 1, ..., 17).$$

5.6 SELF-ASSEMBLING OF IDENTICAL CLUSTERS

Swapping between cations or anions in $A_xB_{1-x}C_yD_{1-y}$ regular solutions can (as shown in Section 5.1) result in the changes of the numbers of bonds (pairs of the nearest neighbors) in accordance with the reaction between bonds $nAC + nBD \rightarrow nAD + nBC$ ($n = 1,...,z_1$) or vice versa. The number n in the reaction $nAC + nBD \rightarrow nAD + nBC$ depends on the nearest neighbors of cations or anions participating in the exchange. If $A_xB_{1-x}C_yD_{1-y}$ regular solution is "non-ideal", i.e., the quantity $\mu_{AC}^0 - \mu_{AD}^0 - \mu_{BC}^0 + \mu_{BD}^0$ is not equal to zero, the nonrandom distribution of cations and anions is preferable. Such distribution reveals a clustering with the primary formation of AC and BD or AD and BC bonds. The primary formation of the pairs of bonds can lead to self-assembling of identical clusters. The self-assembling conditions of identical clusters in $A_xB_{1-x}C_yD_{1-y}$ regular solutions with the zinc blende structure are considered here. The numbers of cations and anions, as usual, correspond to the condition of electroneutrality $N_A + N_B = N_C + N_D = N$. Atoms A and C, accordingly, are chosen in the dilute and ultra-dilute limits ($N_A \ll N_B, N_C \ll N_A$). Moreover, it is supposed that $\mu_{AC}^0 - \mu_{AD}^0 - \mu_{BC}^0 + \mu_{BD}^0 < 0$ and it is also assumed that all interaction parameters between the constituent compounds are equal to zero. Accordingly, the Helmholtz free energy of $A_xB_{1-x}C_yD_{1-y}$ regular solution is the function of the sum of the free energies of the constituent compounds and configurational entropy term given as $F = F_{AC}^0 + F_{AD}^0 + F_{BC}^0 + F_{BD}^0 - TS$. The formation of 1A1C, 2A1C, 3A1C, and 4A1C clusters is more probable at the considered concentration condition than the formation of 1A2C, 1A3C, and 1A4C clusters. The formation of 2A1C and 1A2C, 3A1C and 1A3C, and 4A1C and 1A4C clusters changes the numbers of bonds equally, but the formation 1A2C, 1A3C, and 1A4C clusters decreases the configurational entropy more significantly under the chosen composition condition. The larger clusters than 1A1C, 2A1C, 3A1C, and 4A1C should also decrease considerably the configurational entropy. Thus, $A_xB_{1-x}C_yD_{1-y}$ regular solutions with 1A1C, 2A1C, 3A1C, and 4A1C clusters are taken into account only. The clustering degrees are represented by the 1A1C, 2A1C, 3A1C, and 4A1C cluster parameters, having the form: $\alpha = \frac{N_{1A1C}}{N_C}$, $\beta = \frac{N_{2A1C}}{N_C}$, $\gamma = \frac{N_{3A1C}}{N_C}$, and $\delta = \frac{N_{4A1C}}{N_C}$, respectively, $0 \leq \alpha, \beta, \gamma, \delta \leq 1$, $0 \leq \alpha + \beta + \gamma + \delta \leq 1$, and N_{1A1C} is the number of 1A1C clusters. Due to the dilute and ultra-dilute limits for the numbers of atoms A and C, respectively, the numbers of bonds are written as:

$$N_{AC} = (\alpha + 2\beta + 3\gamma + 4\delta)N_C,$$

$$N_{AD} = z_1 N_A - (\alpha + 2\beta + 3\gamma + 4\delta)N_C,$$

$$N_{BC} = (z_1 - \alpha - 2\beta - 3\gamma - 4\delta)N_C,$$

$$N_{BD} = z_1(N - N_A) - (z_1 - \alpha - 2\beta - 3\gamma - 4\gamma)N_C.$$

The concentrations of bonds are:

$$x_{AC} = \frac{\alpha + 2\beta + 3\gamma + 4\delta}{z_1} y, \quad x_{AD} = x - \frac{\alpha + 2\beta + 3\gamma + 4\delta}{z_1} y,$$

$$x_{BC} = y - \frac{\alpha + 2\beta + 3\gamma + 4\delta}{z_1}, \quad x_{BD} = 1 - x - y + \frac{\alpha + 2\beta + 3\gamma + 4\delta}{z_1} y.$$

The sum of the free energies of the constituent compounds per mole:

$$f^0 = \frac{\alpha + 2\beta + 3\gamma + 4\delta}{z_1} (\mu_{AC}^0 - \mu_{AD}^0 - \mu_{BC}^0 + \mu_{BD}^0) y + \mu_{AD}^0 x + \mu_{BC}^0 y$$
$$+ \mu_{BD}^0 (1 - x - y).$$

The configurational entropy is written as:

$$S = k_B \ln \Omega_1 \Omega_2 \Omega_3,$$

where Ω_1 is the number of configurations of isolated cations with the fixed arrangement of atoms C given by:

$$\Omega_1 = \alpha_{1A1C} \alpha_{2A1C} \alpha_{3A1C} \frac{(N - \alpha N_C - 2\beta N_C - 3\gamma N_C - 4\delta N_C)!}{(N_A - \alpha N_C - 2\beta N_C - 3\gamma N_C - 4\delta N_C)!(N - N_A)!},$$

$\alpha_{1A1C} = 4$, $\alpha_{2A1C} = 6$, and $\alpha_{3A1C} = 4$ are the numbers of the arrangements of $1A1C$, $2A1C$, and $3A1C$ clusters in tetrahedral cell with the central atom C, Ω_2 is the number of arrangements of clusters with the fixed distribution of atoms C, and Ω_3 is the number of configurations of anions written, respectively, as:

$$\Omega_2 = \frac{[(4\alpha + 6\beta + 4\gamma + \delta)N_C]!}{[(\alpha N_C)!]^4 [(\beta N_C)!]^6 [(\gamma N_C)!]^4 (\delta N_C)!},$$

$$\Omega_3 = \frac{N!}{N_C!(N - N_C)!}.$$

The number of configurations Ω_2 is insignificant in comparison with the numbers Ω_1 and Ω_3. Therefore, the number Ω_2 need not be taken into consideration.

The distribution of anions is assumed random due to the ultra-dilute limit for atoms C. Finally, the molar configurational entropy is:

$$s = - R[x - (\alpha + 2\beta + 3\gamma + 4\delta)y]\ln \frac{x - (\alpha + 2\beta + 3\gamma + 4\delta)y}{1 - (\alpha + 2\beta + 3\gamma + 4\delta)y}$$

$$- R(1 - x)\ln \frac{1 - x}{1 - (\alpha + 2\beta + 3\gamma + 4\delta)y} - 4R\alpha y \ln \frac{\alpha}{4\alpha + 6\beta + 4\gamma + \delta}$$

$$- 6R\beta y \ln \frac{\beta}{4\alpha + 6\beta + 4\gamma + \delta} - 4R\gamma y \ln \frac{\gamma}{4\alpha + 6\beta + 4\gamma + \delta}$$

$$- R\delta y \ln \frac{\delta}{4\alpha + 6\beta + 4\gamma + \delta} - R[y \ln y + (1 - y)\ln(1 - y)].$$

The values of cluster parameters are calculated minimizing the Helmholtz free energy $f = f^0 - Ts$:

$$\frac{\partial f}{\partial \alpha} = \frac{1}{z_1}(\mu_{AC}^0 - \mu_{AD}^0 - \mu_{BC}^0 + \mu_{BD}^0) - RT \ln \frac{x - (4\alpha + 3\beta + 2\gamma + \delta)y}{1 - (4\alpha + 3\beta + 2\gamma + \delta)y} = 0,$$

$$\frac{\partial f}{\partial \beta} = \frac{2}{z_1}(\mu_{AC}^0 - \mu_{AD}^0 - \mu_{BC}^0 + \mu_{BD}^0) - 2RT \ln \frac{x - (4\alpha + 3\beta + 2\gamma + \delta)y}{1 - (4\alpha + 3\beta + 2\gamma + \delta)y} = 0,$$

$$\frac{\partial f}{\partial \gamma} = \frac{3}{z_1}(\mu_{AC}^0 - \mu_{AD}^0 - \mu_{BC}^0 + \mu_{BD}^0) - 3RT \ln \frac{x - (4\alpha + 3\beta + 2\gamma + \delta)y}{1 - (4\alpha + 3\beta + 2\gamma + \delta)y} = 0,$$

$$\frac{\partial f}{\partial \delta} = (\mu_{AC}^0 - \mu_{AD}^0 - \mu_{BC}^0 + \mu_{BD}^0) - 4RT \ln \frac{x - (4\alpha + 3\beta + 2\gamma + \delta)y}{1 - (4\alpha + 3\beta + 2\gamma + \delta)y} = 0.$$

These equations are equivalent. Therefore, it is sufficient to analyze one of them:

$$0 \leq \alpha, \beta, \gamma, \delta \leq 1, 0 \leq \alpha + \beta + \gamma + \delta \leq 1,$$

$$\mu_{AC}^0 - \mu_{AD}^0 - \mu_{BC}^0 + \mu_{BD}^0 - 4RT \ln \frac{x - (4\alpha + 3\beta + 2\gamma + \delta)y}{1 - (4\alpha + 3\beta + 2\gamma + \delta)y} = 0$$

or

$$T = \frac{\mu_{AC}^0 - \mu_{AD}^0 - \mu_{BC}^0 + \mu_{BD}^0}{4R \ln \dfrac{x - (\alpha + 2\beta + 3\gamma + 4\delta)y}{1 - (\alpha + 2\beta + 3\gamma + 4\delta)y}}.$$

The temperatures of the occurrence of the clusters are given by:

$$T_{1(1A1C)} \approx T_{1(2A1C)} \approx T_{1(3A1C)} \approx T_{1(4A1C)} = \frac{\mu_{AC}^0 - \mu_{AD}^0 - \mu_{BC}^0 + \mu_{BD}^0}{4R \ln x},$$

$$\text{but } T_{1(1A1C)} > T_{1(2A1C)} > T_{1(3A1C)} > T_{1(4A1C)}.$$

However, the differences between the temperatures T_1 are insignificant. The temperatures, when the same finite portion ε of atoms C are in the same clusters, are:

$$T_{2(1A1C)} = \frac{\mu_{AC}^0 - \mu_{AD}^0 - \mu_{BC}^0 + \mu_{BD}^0}{4R \ln \frac{x - \varepsilon y}{1 - \varepsilon y}},$$

$$T_{2(2A1C)} = \frac{\mu_{AC}^0 - \mu_{AD}^0 - \mu_{BC}^0 + \mu_{BD}^0}{4R \ln \frac{x - 2\varepsilon y}{1 - 2\varepsilon y}},$$

$$T_{2(3A1C)} = \frac{\mu_{AC}^0 - \mu_{AD}^0 - \mu_{BC}^0 + \mu_{BD}^0}{4R \ln \frac{x - 3\varepsilon y}{1 - 3\varepsilon y}},$$

$$T_{2(4A1C)} = \frac{\mu^0_{AC} - \mu^0_{AD} - \mu^0_{BC} + \mu^0_{BD}}{4R \ln \frac{x - 4\varepsilon y}{1 - 4\varepsilon y}},$$

where $0 < \varepsilon < 1$ and $T_{2(1A1C)} > T_{2(2A1C)} > T_{2(3A1C)} > T_{2(4A1C)}$. The differences between the temperatures T_2 increase with increasing the portion of atoms C in clusters. The temperatures of the completion of the self-assembling process when all atoms C are in the same clusters ($\varepsilon = 1$) are:

$$T_{3(1A1C)} = \frac{\mu^0_{AC} - \mu^0_{AD} - \mu^0_{BC} + \mu^0_{BD}}{4R \ln \frac{x - y}{1 - y}},$$

$$T_{3(2A1C)} = \frac{\mu^0_{AC} - \mu^0_{AD} - \mu^0_{BC} + \mu^0_{BD}}{4R \ln \frac{x - 2y}{1 - 2y}},$$

$$T_{3(3A1C)} = \frac{\mu^0_{AC} - \mu^0_{AD} - \mu^0_{BC} + \mu^0_{BD}}{4R \ln \frac{x - 3y}{1 - 3y}},$$

$$T_{3(4A1C)} = \frac{\mu^0_{AC} - \mu^0_{AD} - \mu^0_{BC} + \mu^0_{BD}}{4R \ln \frac{x - 4y}{1 - 4y}}$$

and $T_{3(1A1C)} > T_{3(2A1C)} > T_{3(3A1C)} > T_{3(4A1C)}$. The differences between the temperatures reach the maximal values when all atoms C are in the clusters. In the case when atoms C can be in all types of clusters and under the condition $0 < \alpha + \beta + \gamma + \delta \leq 1$, the maximal temperature should be reached if atoms C are in $4A1C$ clusters.

The obtained results demonstrate that self-assembling of $4A1C$ identical clusters in $A_xB_{1-x}C_yD_{1-y}$ regular solutions with $x \ll 1$ and $y \ll x$ is preferable from the thermodynamics standpoint if the condition $\mu^0_{AC} - \mu^0_{AD} - \mu^0_{BC} + \mu^0_{BD} < 0$ is fulfilled.

As was shown in Section 5.3.3, the quantity $\mu^0_{AC} - \mu^0_{AD} - \mu^0_{BC} + \mu^0_{BD}$ can be estimated as:

$$\mu^0_{AC} - \mu^0_{AD} - \mu^0_{BC} + \mu^0_{BD} = \sum_{i,j=1}^{2} (-1)^i (-1)^j$$

$$\times \left[h^{0f}_{ij} (298.15 \text{ K}) - Ts_{ij}(298.15 \text{ K}) + \int_{298.15}^{T} c^p_{ij} dT - T \int_{298.15}^{T} \frac{c^p_{ij}}{T} dT \right]$$

where $i = A$, B; $j = C$, D; h^{0f}_{ij} and s_{ij} are the standard enthalpy of formation and standard entropy of ij-th compound, respectively, and c^p_{ij} is the heat capacity at the constant pressure. The standard enthalpies of formation significantly contribute to the quantity

$\mu^0_{AC} - \mu^0_{AD} - \mu^0_{BC} + \mu^0_{BD}$. Therefore, in many cases, the quantity $h^{0f}_{AC}(298.15 \text{ K}) - h^{0f}_{AD}(298.15 \text{ K}) - h^{0f}_{BC}(298.15 \text{ K}) + h^{0f}_{BD}(298.15 \text{ K})$ can be used instead of the quantity $\mu^0_{AC} - \mu^0_{AD} - \mu^0_{BC} + \mu^0_{BD}$.

The enthalpy of formation of AlN is extremely large among the enthalpies of formation of the $A^{III}B^V$ semiconductor compounds. Accordingly, the formation of AlN bonds in $Al_xA^{III}_{1-x}N_yB^V_{1-y}$ semiconductor alloys is very thermodynamically preferred. The almost complete absence of GaN bonds in GaAs-rich $Al_xGa_{1-x}N_yAs_{1-y}$ alloys with $x \gg y$ was established experimentally [1]. Redistribution of Al and Ga atoms on the cation sublattice sites in $Al_xGa_{1-x}N_yAs_{1-y}$ alloys cannot lead to the significant variation of the strain energy, since the lattice parameters of AlN and GaN as well as the lattice parameters of AlAs and GaAs are close. Therefore, the considerable value of the quantity $\mu^0_{AlN} - \mu^0_{AlAs} - \mu^0_{GaN} + \mu^0_{GaAs}$ is the cause of the insignificant formation of GaN bonds in the studied $Al_xGa_{1-x}N_yAs_{1-y}$ alloys. Accordingly, the self-assembling conditions of 1N4Al clusters in GaAs-rich $Al_xGa_{1-x}N_yAs_{1-y}$ alloys can be considered by the presented approach. The same approach can be also used to estimate the self-assembling conditions of $1N4B^V$ clusters in GaB^V-rich $Al_xGa_{1-x}N_yB^V_{1-y}$ (B^V = P, Sb) alloys with $x \gg y$ and $1B^V4Ga$ clusters in AlN-rich $Ga_xAl_{1-x}B^V_yN_{1-y}$ alloys also with $x \gg y$, since the variation of the strain energy as a result of redistribution of Al and Ga atoms in these alloys is also insignificant.

Many $A^{II}B^{VI}$ semiconductor compounds due to their strong ionicity have a large value of the enthalpy of formation. Therefore, a lot of $A^{II}_xB^{II}_{1-x}C^{VI}_yD^{VI}_{1-y}$ semiconductor alloys have a sufficiently large value of quantity $\mu^0_{AC} - \mu^0_{AD} - \mu^0_{BC} + \mu^0_{BD}$. Self-assembling of identical tetrahedral clusters in these alloys should be thermodynamically favorable. However, the redistribution of cations or anions should change significantly the strain energy in $A^{II}_xB^{II}_{1-x}C^{VI}_yD^{VI}_{1-y}$ alloys. This has to be taken into account when considering the self-assembling conditions. The strain energy in such alloys can be described in the framework of the valence force field model presented in Chapter 6.

Reference

[1] T. Geppert, J. Wagner, K. Köhler, P. Ganser, M. Maier, Preferential formation of Al–N bonds in low N-content AlGaAsN, Appl. Phys. Lett. 80 (12) (2002) 2081–2083.

6

Valence Force Field Model and Its Applications

The valence force field model is the most frequently used model for considering semiconductor alloys when taking into account the distortions of the crystal lattice on the microscopic scale. Initially this model was developed for diamond and the elemental semiconductors Si, Ge, and gray Sn with the diamond structure [1]. Later, the model was extended to the binary semiconductor compounds with the zinc blende structure [2]. Semiconductors with the wurtzite structure also can be considered with the valence force field model [2,3]. This is a consequence of the facts that the tetrahedrally coordinated zinc blende and wurtzite structures are similar and the distances between the nearest neighbors in the same compounds with these structures are close.

Even the first application of the valence force field model to the description of $In_xGa_{1-x}As$ semiconductor alloys demonstrated the significant progress in the calculation of the distortions of the crystal lattice on the microscopic scale [4]. The theoretically estimated distortions were found sufficiently close to the experimental results. The further studies of many III–V and II–VI ternary semiconductor alloys with the zinc blende and wurtzite structures showed that the basic peculiarities of the crystal structure distortions are very similar. Thus, the valence force field model is up until now the basic model for the consideration of the crystal lattice distortions and other characteristics of semiconductor alloys.

6.1 CRYSTAL STRUCTURE OF TERNARY ALLOYS OF BINARY COMPOUNDS

The crystal lattice of $A_xB_{1-x}C$ regular solutions of binary compounds is assumed to be geometrically undistorted with the lattice parameter obeying Vegard's law, which determines the lattice parameter and distances between the nearest neighbors, correspondingly, as:

$$a = a_{AC}x + a_{BC}(1 - x),$$

$$r_{AC} = R_{AC} - (R_{AC} - R_{BC})(1 - x),$$

$$r_{BC} = R_{BC} + (R_{AC} - R_{BC})x$$

$$\text{and } r_{AC} = r_{BC},$$

where a_{AC} and R_{AC} are, consequently, the lattice parameter and the distances between the nearest neighbors in the binary compound AC. Thus, the distances between the nearest neighbors or the lengths of bonds in the molecular regular solutions are assumed equal. Moreover, the angles between bonds are assumed to be undistorted. These assumptions about the crystal structure of the regular solution correspond to the virtual crystal approximation in solid-state physics.

The experimental results of the distortions of the crystal structure of the zinc blende $In_xGa_{1-x}As$ semiconductor alloys obtained by using the extended X-ray absorption fine-structure method [5] gave unexpected information on the structural distortions. The crystal structure of $In_xGa_{1-x}As$ alloys studied does not correspond to the virtual crystal approximation. In fact, the main peculiarities of the crystal structure have three aspects: (1) atoms forming the mixed sublattice randomly occupy the lattice sites, (2) the mixed sublattice is slightly distorted and obeys Vegard's law, and (3) the distances between the nearest neighbors depend weakly on the composition of $In_xGa_{1-x}As$ alloy.

Later, the extended X-ray absorption fine-structure method was used to study the crystal structures of many $A_x^{III}B_{1-x}^{III}C^V, A^{III}B_x^VC_{1-x}^V, A_x^{II}B_{1-x}^{II}C^{VI}$, and $A^{II}B_x^{VI}C_{1-x}^{VI}$ semiconductor alloys with the zinc blende structure [6,7]. The results of all these studies demonstrated the same peculiarities of the crystal structure distortions as were obtained for $In_xGa_{1-x}As$ alloys. The extended X-ray absorption fine-structure studies of wurtzite $A_xB_{1-x}C$ semiconductor alloys [8] showed that the basic peculiarities of the crystal structure of the wurtzite ternary alloys are very similar to the results obtained for the zinc blende ternary alloys. In these results the same three points may be distinguished: (1) the atoms of the mixed sublattice are distributed randomly on their sublattice sites; (2) the mixed sublattice is distorted slightly, but the other sublattice is distorted strongly; and (3) the distances of the nearest atoms depend slightly on the composition of the alloys. In accordance with these results, the distances between the nearest neighbors in the $A_xB_{1-x}C$ semiconductor alloys with the zinc blende and wurtzite structures can be approximately expressed as:

$$r_{AC} \approx R_{AC} - \frac{1}{4}(R_{AC} - R_{BC})(1 - x),$$

$$r_{BC} \approx R_{BC} + \frac{1}{4}(R_{AC} - R_{BC})x,$$

where R_{AC} is the distance between the nearest neighbors in the compound AC.

6.2 VALENCE FORCE FIELD MODEL

The distorted crystal structure of the ternary semiconductor alloys can be described using the valence force field model developed for the consideration of the strained state of semiconductors with the diamond, zinc blende, and wurtzite structures [2,3]. In the valence force field model, the strained state of elemental semiconductors or binary compounds is represented by using two types of elastic constants. The constants of the first type, called the bond-stretching constants, are the elastic constants of bonds between the nearest neighbors. The constants of the second type, named the bond-bending elastic constants, are the elastic constants of the angles between the bonds.

The elastic constants of elemental semiconductors with the diamond structure are given by:

$$\alpha = \frac{8}{\sqrt{3}}(C_{11} + C_{12})R, \quad \beta = \frac{8}{\sqrt{3}}(C_{11} - C_{12})R,$$

where α and β are the bond-stretching and bond-bending elastic constants, respectively, R is the distance between the nearest neighbors in the undistorted crystal, and C_{ij} are the stiffness coefficients.

The elastic constants of binary semiconductor compounds with the zinc blende structure are obtained from the system of equations written as:

$$3\alpha - 18.5C_{11}R - 11.5C_{12}R - 29\beta = 0,$$

$$0.433(\alpha + \beta) - 2.57[(C_{11} + C_{12})R + 1.73\beta]$$

$$- \frac{[0.433(\alpha - \beta) - 5.55(C_{11}R - C_{12}R - 1.73\beta)]^2}{0.433(\alpha + \beta) - 5.02(C_{11}R - C_{12}R - 1.73\beta)} - C_{44}R = 0,$$

where α and β are the bond-stretching and bond-bending elastic constants, R is the distance between the nearest neighbors in the undistorted crystal, and C_{ij} are the stiffness coefficients. As established in Ref. [3], the bond-stretching and bond-bending elastic constants of the same binary semiconductor compound with the zinc blende and wurtzite structures should differ insignificantly. Moreover, the stiffness coefficients of the semiconductor with the zinc blende structure can be obtained by using the stiffness coefficients of the same semiconductor with the wurtzite structure, and vice versa. Thus, the consideration of semiconductor alloys with constituent compounds having the zinc blende and wurtzite structures in their stable state may be fulfilled by using the valence force field model.

The deformation energy of the primitive cell of elemental semi-conductors or binary compounds consisting of two atoms described using the valence force field model is:

$$u = \frac{3}{8R^2}\left[\alpha\sum_{i=1}^{4}\left(R^2 - r_i^2\right)^2 + \beta\sum_{k=1}^{2}\sum_{i,j>i}\left(R^2\cos\varphi_0 - r_{ik}r_{jk}\cos\varphi_{ik-jk}\right)^2\right],$$

where R is the distance between the nearest neighbors in the undistorted crystal, r_i is the length of i-th bond around the first atom in the primitive cell, r_{ik} and r_{jk} are the lengths of different bonds situated around atom k, and $\varphi_0 = 109.47°$ (cos $\varphi_0 = -1/3$) and φ_{ik-jk} are the angles between the ik-th and jk-th bonds in the undistorted and distorted crystals, respectively. Thus, the deformation energies of a bond and an angle between bonds, respectively, are:

$$u_b = \frac{3\alpha\left(R^2 - r^2\right)^2}{8R^2}, \tag{6.2.1}$$

$$u_a = \frac{3\beta\left(R^2\cos\varphi_0 - r_1r_2\cos\varphi\right)^2}{8R^2}, \tag{6.2.2}$$

where R is the distance between the nearest neighbors in the undistorted crystal, and r, r_1, and r_2 are the distances between the nearest neighbors in the distorted crystal.

Now let us consider the crystal structure and the internal strain energy of ternary alloys of binary compounds with the zinc blende structure. In accordance with the experimental results obtained by the extended X-ray absorption fine-structure method described in Section 6.1, it is supposed that the mixed sublattice is geometrically undistorted. A tetrahedral unit cell with four atoms in the vertices and the atom in its center is a basic unit in the description of the lattice distortions. $A_xB_{1-x}C$ ternary alloys in which the lattice parameter of the compound AC is not less than the lattice parameter of the compound BC are considered. The mixed sublattice may be cationic or anionic, and thus the alloy under consideration may be both $A_xB_{1-x}C$ and AB_xC_{1-x} alloys. Atoms in the vertices of tetrahedral cells are atoms of the mixed sublattice and a central atom is an atom of another sublattice. There are five types of tetrahedral cells—$4A1C$, $3A1B1C$, $2A2B1C$, $1A3B1C$, and $4B1C$—in $A_xB_{1-x}C$ alloys. The tetrahedral cells are shown in Figure 6.1(a−d).

The size of all tetrahedral cells is the same according to the supposition of the undistorted mixed sublattice. Thus, the internal strain energy of tetrahedral cells is a function of the composition of the alloy and the displacement of the atom C from the geometrical center of the tetrahedral

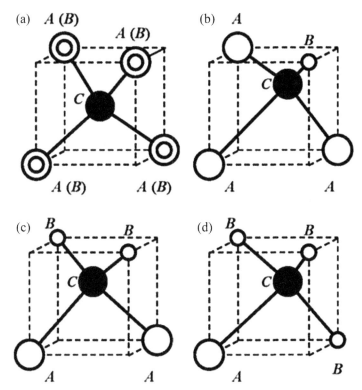

FIGURE 6.1 (a–d) Tetrahedral cells 4A1C (a), 3A1B1C (b), 2A2B1C (c), 1A3B1C (d), and 4B1C (a) in $A_xB_{1-x}C$ alloys.

cell. The displacements of the atoms C are calculated by minimizing the internal strain energy of 3A1B1C, 2A2B1C, 1A3B1C tetrahedral cells, which is $\frac{du_i}{dw_i} = 0$, where u_i ($i = 3A1B1C, 2A2B1C, 1A3B1C$) and w_i are the internal strain energy and the displacement of the central atom in the tetrahedral cell, respectively.

The displacements of the atoms C and the internal strain energies of tetrahedral cells of $A_xB_{1-x}C$ alloys under the condition when the displacement is significantly smaller than the lattice parameters of the constituent compounds are:

$$w_{4A1C} = 0, \tag{6.2.3}$$

$$u_{4A1C} = 2(3\alpha_{AC} + \beta_{AC})\Delta R^2(1 - x)^2, \tag{6.2.4}$$

$$w_{3A1B1C} = \frac{3[\alpha_{AC} + (\alpha_{BC} - \alpha_{AC})x] + \frac{5\beta_{AC} + \beta_{BC}}{2} - (3\beta_{AC} + \beta_{BC})x}{\alpha_{AC} + 3\alpha_{BC} + 3\beta_{AC} + \beta_{BC}}\Delta R, \tag{6.2.5}$$

$$u_{3A1B1C} = \frac{3}{2}\left\{3\alpha_{AC}\left[\Delta R(1-x) - \frac{w_{3A1B}}{3}\right]^2 + \alpha_{BC}(\Delta Rx - w_{3A1B})^2\right\}$$

$$+ \frac{\beta_{AC} + \beta_{BC}}{8}\left[\Delta R(1-2x) - 2w_{3A1B}\right] + \beta_{AC}\left[\Delta R(1-x) - w_{3A1B}\right]^2,$$

$$(6.2.6)$$

$$w_{2A2B1C} = \sqrt{3}\,\frac{\alpha_{AC}(1-x) + \alpha_{BC}x - \frac{\beta_{AC}}{3}(1-x) - \frac{\beta_{BC}}{3}x}{\alpha_{AC} + \alpha_{BC} + \beta_{AC} + \beta_{BC}}\Delta R, \qquad (6.2.7)$$

$$u_{2A2B1C} = 3\left\{\alpha_{AC}\left[\Delta R(1-x) - \frac{w_{2A2B}}{\sqrt{3}}\right]^2 + \alpha_{BC}\left(\Delta Rx - \frac{w_{2A2B}}{\sqrt{3}}\right)^2\right\}$$

$$+ \frac{\beta_{AC}}{3}\left[\Delta R(1-x) + \sqrt{3}w_{2A2B}\right] + \frac{\beta_{BC}}{3}\left(\Delta Rx + \sqrt{3}w_{2A2B}\right)^2$$

$$+ \frac{\beta_{AC} + \beta_{BC}}{6}\Delta R^2(1-2x),$$

$$(6.2.8)$$

$$w_{1A3B1C} = \frac{3[\alpha_{AC} + (\alpha_{BC} - \alpha_{AC})x] + (\beta_{AC} - \beta_{BC})x - \frac{\beta_{AC}+\beta_{BC}}{2}}{3\alpha_{AC} + \alpha_{BC} + \beta_{AC} + 3\beta_{BC}}\Delta R, \quad (6.2.9)$$

$$u_{1A3B1C} = \frac{3}{2}\left\{\alpha_{AC}[\Delta R(1-x) - w_{1A3B}]^2 + 3\alpha_{BC}\left(\Delta Rx - \frac{w_{1A3B}}{3}\right)^2\right\}$$

$$+ \frac{\beta_{AC} + \beta_{BC}}{8}\left[\Delta R(1-2x) + 2w_{1A3B}\right] + \beta_{BC}(\Delta Rx + w_{1A3B})^2,$$

$$(6.2.10)$$

$$w_{4B1C} = 0, \qquad (6.2.11)$$

$$u_{4B1C} = 2(3\alpha_{BC} + \beta_{BC})\Delta R^2 x^2, \qquad (6.2.12)$$

where u_{4A1C} and w_{4A1C} are the internal strain energy or the deformation energy of 4A1C tetrahedral cell and the displacement of the central atom C in 4A1C tetrahedral cell, respectively; $\Delta R = R_{AC} - R_{BC}$, $R_{AC} > R_{BC}$ and R_{AC} is the distance between A and C atoms in the unstrained compound AC; α_{AC} and β_{AC} are the bond-stretching and bond-bending elastic constants of the compound AC, respectively; and w_{3A1B1C} is the displacement of the atom C in 3A1B1C tetrahedral cell from its geometrical center. The bond-bending elastic constant between AC and BC bonds is assumed to be equal to the arithmetical average of the bond-bending elastic constants of AC and BC compounds:

$$\beta_{AC-BC} = \frac{\beta_{AC} + \beta_{BC}}{2}. \qquad (6.2.13)$$

This bond-bending elastic constant was used to obtain the formulas of the displacements and deformation energies. The average deformation energy of a tetrahedral cell in $A_xB_{1-x}C$ alloy with the random distribution of atoms in the mixed sublattice is:

$$\bar{u} = u_{4A1C}x^4 + 4u_{3A1B1C}x^3(1-x) + 6u_{2A2B1C}x^2(1-x)^2$$
$$+ 4u_{1A3B1C}x(1-x)^3 + u_{4B1C}(1-x)^4. \tag{6.2.14}$$

The average distances between the nearest neighbors in such an alloy are:

$$\overline{r_{AC}} = r_{AC}^{4A1C}x^3 + 3r_{AC}^{3A1B1C}x^2(1-x) + 3r_{AC}^{2A2B1C}x(1-x)^2 + r_{AC}^{3A1B1C}(1-x)^3,$$

$$\overline{r_{BC}} = r_{BC}^{3A1B1C}x^3 + 3r_{BC}^{2A2B1C}x^2(1-x) + 3r_{BC}^{1A3B1C}x(1-x)^2 + r_{BC}^{4B1C}(1-x)^3,$$

where r_{AC}^{4A1C} is the distance between A and C atoms in the 4A1C tetrahedral cell. The average distances between the nearest atoms in $A_x^{III}B_{1-x}^{III}C^V$ and $A_x^{II}B_{1-x}^{II}C^{VI}$ semiconductor alloys can be approximately expressed as:

$$\overline{r_{AC}} \approx R_{AC} - \frac{1}{2}(R_{AC} - R_{BC})(1-x) \text{ and}$$

$$\overline{r_{BC}} \approx R_{BC} + \frac{1}{2}(R_{AC} - R_{BC})x,$$

where R_{AC} is the distance between the nearest neighbors in the compound AC. The valence force field model provides the significant progress in the description of the crystal structure of $A_x^{III}B_{1-x}^{III}C^V$ and $A_x^{II}B_{1-x}^{II}C^{VI}$ alloys in comparison with the regular solution model (the virtual crystal approximation), since the average distances between the nearest neighbors described in the regular solution model are:

$$\overline{r_{AC}^{VCA}} = R_{AC} - (R_{AC} - R_{BC})(1-x),$$

$$\overline{r_{BC}^{VCA}} = R_{BC} + (R_{AC} - R_{BC})x,$$

$$\overline{r_{AC}^{VCA}} = \overline{r_{BC}^{VCA}}.$$

The deformation energies of the tetrahedral cells of $In_xGa_{1-x}P$, $In_xGa_{1-x}As$, $GaSb_xAs_{1-x}$, and $ZnSe_xS_{1-x}$ alloys and their average deformation energies for several compositions estimated by using formulas (6.2.3–6.2.12) are shown in Tables 6.1–6.4. The deformation energies of bonds and angles are denoted by (α) and (β), respectively. The average deformation energies of tetrahedral cells $\bar{u}$, the average deformation energies of bonds $\overline{u(\alpha)}$, and the average deformation energies of angles $\overline{u(\beta)}$ are also shown.

The average deformation energies of the tetrahedral cells of these alloys are close to the symmetric functions with respect to composition $x = 0.5$ in spite of the different elastic constants of binary constituent compounds.

TABLE 6.1   The deformation energies of the tetrahedral cells of $In_xGa_{1-x}P$ alloys and their average deformation energies

kJ mol^{-1}	$x = 0$	$x = 0.25$	$x = 0.50$	$x = 0.75$	$x = 1$
u_{4In1P}	53.427	30.052	13.357	3.339	0
$u_{4In1P}(\alpha)$	50.964	28.667	12.741	3.185	0
$u_{4In1P}(\beta)$	2.463	1.385	0.616	0.154	0
$u_{3In1Ga1P}$	29.507	13.388	4.180	1.883	6.497
$u_{3In1Ga1P}(\alpha)$	29.331	13.112	3.570	0.708	4.525
$u_{3In1Ga1P}(\beta)$	0.177	0.276	0.610	1.175	1.972
$u_{2In2Ga1P}$	17.162	6.887	3.705	7.617	18.621
$u_{2In2Ga1P}(\alpha)$	13.814	3.858	0.577	3.972	14.041
$u_{2In2Ga1P}(\beta)$	3.348	3.029	3.128	3.645	4.580
$u_{1In3Ga1P}$	6.426	3.056	6.990	18.229	36.773
$u_{1In3Ga1P}(\alpha)$	3.905	0.537	4.004	14.308	31.449
$u_{1In3Ga1P}(\beta)$	2.521	2.519	2.986	3.921	5.324
u_{4Ga1P}	0	3.760	15.038	33.836	60.153
$u_{4Ga1P}(\alpha)$	0	3.502	14.008	31.518	56.032
$u_{4Ga1P}(\beta)$	0	0.258	1.030	2.318	4.121
$\bar{u}$	0	4.676	5.957	4.444	0
$\overline{u(\alpha)}$	0	2.875	3.782	2.938	0
$\overline{u(\beta)}$	0	1.801	2.175	1.506	0

Moreover, the deformation energies can be represented with reasonable accuracy as functions $u \approx \alpha x(1 - x)$, where α is a constant. Thus, $A_xB_{1-x}C$ semiconductor alloys can be considered as molecular strictly regular alloys from the internal strain energy standpoint. As seen from Tables 6.1–6.4, the average deformation energies of bonds are significantly larger than the average deformation energies of angles between bonds. Moreover, the deformation energies of $3A1B1C$, $2A2B1C$, and $1A3B1C$ tetrahedral cells of $A_xB_{1-x}C$ alloys ($In_xGa_{1-x}P$, $In_xGa_{1-x}As$, $GaSb_xAs_{1-x}$, and $ZnSe_xS_{1-x}$) are smaller than the average deformation energies of tetrahedral cells with the same composition. It means that there is a tendency to the preferential formation of $3A1B1C$, $2A2B1C$, and $1A3B1C$ tetrahedral cells at compositions close to $x = 0.75$, $x = 0.5$, and $x = 0.25$, respectively, from the internal strain energy standpoint. $A_xB_{1-x}C$ ($x = 0.75, 0.5$, and 0.25) alloy consisting of one-type tetrahedral cells can have the completely ordered long-distance

TABLE 6.2 The deformation energies of the tetrahedral cells of $In_xGa_{1-x}As$ alloys and their average deformation energies

kJ mol^{-1}	$x = 0$	$x = 0.25$	$x = 0.50$	$x = 0.75$	$x = 1$
u_{4In1As}	40.970	23.046	10.243	2.561	0
$u_{4In1As}(\alpha)$	38.941	21.904	9.735	2.434	0
$u_{4In1As}(\beta)$	2.029	1.142	0.507	0.127	0
$u_{3In1Ga1As}$	22.883	10.393	3.350	1.754	5.604
$u_{3In1Ga1As}(\alpha)$	22.734	10.198	2.938	0.953	4.244
$u_{3In1Ga1As}(\beta)$	0.149	0.195	0.412	0.801	1.360
$u_{2In2Ga1As}$	13.477	5.433	2.989	6.143	14.896
$u_{2In2Ga1As}(\alpha)$	10.871	3.046	0.476	3.159	11.097
$u_{2In2Ga1As}(\beta)$	2.606	2.387	2.513	2.984	3.799
$u_{1In3Ga1As}$	5.090	2.445	5.651	14.707	29.615
$u_{1In3Ga1As}(\alpha)$	3.125	0.442	3.227	11.480	25.203
$u_{1In3Ga1As}(\beta)$	1.964	2.003	2.424	3.227	4.412
u_{4Ga1As}	0	3.056	12.224	27.504	48.896
$u_{4Ga1As}(\alpha)$	0	2.850	11.398	25.646	45.594
$u_{4Ga1As}(\beta)$	0	0.206	0.826	1.858	3.302
$\bar{u}$	0	3.722	4.775	3.643	0
$\overline{u(\alpha)}$	0	2.294	3.040	2.477	0
$\overline{u(\beta)}$	0	1.428	1.735	1.166	0

arrangement of A and B atoms named the complete long-range order or the ideal superstructure (in crystallography) or the ideal superlattice (in statistical physics). In such a case, the mixed sublattice consists of two not necessarily geometrically equivalent sublattices a and b. One of these sublattices fills with atoms A, and atoms B belong to another sublattice. In the general case, there is a complete long-range order in $A_xB_{1-x}C$ $(0 < x < 1)$ alloy if all minority atoms situated in the mixed sublattice belong to the same sublattice, a or b. In other words, the long-range order occurs in such an alloy if one sublattice a or b of the mixed sublattice contains more one-type atoms than in the case of the disordered alloy (alloy without superstructure). The degree of the long-range order is described by the long-range order parameter, which is a function of a probability of the arrangement in any one-type atoms sublattice a or b. The long-range order parameter is normally chosen for the sake of

TABLE 6.3   The deformation energies of the tetrahedral cells of $GaSb_xAs_{1-x}$ alloys and their average deformation energies

kJ mol^{-1}	$x = 0$	$x = 0.25$	$x = 0.50$	$x = 0.75$	$x = 1$
u_{4Sb1Ga}	46.897	26.380	11.724	2.931	0
$u_{4Sb1Ga}(\alpha)$	43.724	24.595	10.931	2.733	0
$u_{4Sb1Ga}(\beta)$	3.173	1.785	0.793	0.198	0
$u_{3Sb1As1Ga}$	26.093	11.880	4.059	2.618	7.561
$u_{3Sb1As1Ga}(\alpha)$	25.859	11.646	3.650	1.738	6.045
$u_{3Sb1As1Ga}(\beta)$	0.224	0.234	0.409	0.880	1.516
$u_{2Sb2As1Ga}$	16.095	6.733	3.892	7.571	17.770
$u_{2Sb2As1Ga}(\alpha)$	12.665	3.650	0.696	3.800	12.965
$u_{2Sb2As1Ga}(\beta)$	3.430	3.083	3.196	3.771	4.807
$u_{1Sb3As1Ga}$	6.123	3.014	6.791	17.456	35.008
$u_{1Sb3As1Ga}(\alpha)$	3.725	0.589	3.855	13.528	29.606
$u_{1Sb3As1Ga}(\beta)$	2.398	2.426	2.936	3.928	5.402
u_{4As1Ga}	0	3.640	14.561	32.760	58.246
$u_{4As1Ga}(\alpha)$	0	3.394	13.578	30.550	54.312
$u_{4As1Ga}(\beta)$	0	0.246	0.983	2.213	3.934
$\bar{u}$	0	4.504	5.815	4.575	0
$\overline{u(\alpha)}$	0	2.734	3.658	3.153	0
$\overline{u(\beta)}$	0	1.770	2.157	1.422	0

convenience. The simplest instance of the long-range order parameter is a probability of the arrangement of one type atoms in any sublattice. However, in most cases, the long-range order parameter is chosen so that it is equal to zero for a disordered alloy and equal to one for a completely long-range ordered alloy. In any case, the value of the long-range order parameter of the alloy with superstructure is larger than the value of the long-range order parameter of the disordered alloy.

The tendency to the formation of the superstructure established by using the valence force field model is the important peculiarity of $A_xB_{1-x}C$ semiconductor alloys. Also it is known that all $A_xB_{1-x}C$ lattice mismatched alloys have the tendency to the phase separation from the internal strain energy standpoint. Thus, there are simultaneously two competitive tendencies in such alloys: the tendency to phase separation and the tendency to formation of the superstructure. Accordingly, $A_xB_{1-x}C$ alloys can be in

TABLE 6.4 The deformation energies of the tetrahedral cells of $ZnSe_xS_{1-x}$ alloys and their average deformation energies

kJ mol^{-1}	$x = 0$	$x = 0.25$	$x = 0.50$	$x = 0.75$	$x = 1$
u_{4Se1Zn}	16.617	9.347	4.154	1.038	0
$u_{4Se1Zn}(\alpha)$	15.978	8.987	3.994	0.998	0
$u_{4Se1Zn}(\beta)$	0.639	0.360	0.160	0.040	0
$u_{3Se1S1Zn}$	9.549	4.323	1.396	0.767	2.437
$u_{3Se1S1Zn}(\alpha)$	9.500	4.278	1.308	0.590	2.123
$u_{3Se1S1Zn}(\beta)$	0.049	0.045	0.088	0.177	0.314
$u_{2Se2S1Zn}$	5.299	1.894	0.816	2.066	5.645
$u_{2Se2S1Zn}(\alpha)$	4.542	1.193	0.082	1.211	4.578
$u_{2Se2S1Zn}(\beta)$	0.757	0.701	0.733	0.855	1.067
$u_{1Se3S1Zn}$	1.792	0.625	1.931	5.711	11.965
$u_{1Se3S1Zn}(\alpha)$	1.252	0.069	1.267	4.849	10.813
$u_{1Se3S1Zn}(\beta)$	0.540	0.556	0.664	0.862	1.152
u_{4S1Zn}	0	1.318	5.272	11.862	21.089
$u_{4S1Zn}(\alpha)$	0	1.273	5.091	11.456	20.366
$u_{4S1Zn}(\beta)$	0	0.045	0.181	0.406	0.723
$\bar{u}$	0	1.319	1.727	1.402	0
$\overline{u(\alpha)}$	0	0.919	1.243	1.092	0
$\overline{u(\beta)}$	0	0.400	0.484	0.310	0

the stable, metastable, or unstable state with respect to phase separation and formation of the superstructure.

As an example, the deformation energies of the tetrahedral cells and the average deformation energies in $In_xGa_{1-x}As$ alloys obtained by using the valence force field model and by using the virtual crystal approximation are shown, correspondingly, in Figures 6.2 and 6.3.

The deformation energies of 3In1Ga1As, 2In2Ga1As, and 1In3Ga1As tetrahedral cells estimated by using the virtual crystal approximation are significantly greater than the deformation energies obtained by the valence force field model. This happens because the bond-stretching constants are considerably larger than the bond-bending constants. Moreover, the deformation energies of 3In1Ga1As, 2In2Ga1As, and 1In3Ga1As tetrahedral cells obtained by using the virtual crystal approximation are equal to the average internal strain energies of

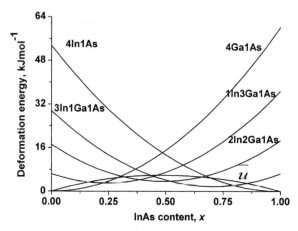

FIGURE 6.2 The deformation energies of tetrahedral cells and average deformation energies in $In_xGa_{1-x}As$ alloys obtained by using the valence force field model.

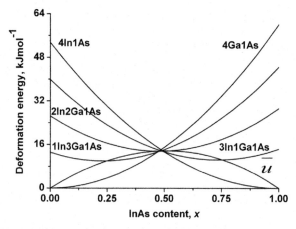

FIGURE 6.3 The deformation energies of tetrahedral cells and average deformation energies in $In_xGa_{1-x}As$ alloys obtained by using virtual crystal approximation.

tetrahedral cells with the composition corresponding to the same type of the tetrahedral cell. The deformation energies of tetrahedral cells estimated by the virtual crystal approximation exclude the formation of the long-range order, since the free energy of the ordered alloy is larger than the strain energy of the disordered alloy due to the smaller configurational entropy. The same is true for all ternary alloys of two binary compounds when describing the internal strain energies in the virtual crystal approximation.

6.3 SUPERSTRUCTURES

There are two types of the ideal superstructures that can be formed only by $3A1B1C$ or $1A3B1C$ tetrahedral cells in $A_xB_{1-x}C$ alloys with the zinc blende structure. The first type, shown in Figure 6.4(a) is the luzonite-type superstructure having the crystal structure similar to the crystal structure of luzonite Cu_3AsS_4.

In the completely ordered state, the ideal luzonite-type superstructure has one of four simple cubic sublattices constituting the face-centered cubic mixed sublattice filled with one-type atoms. The other three simple cubic sublattices entering in the same face-centered cubic mixed sublattice are completed with atoms of another type. The second type of the superstructure shown in Figure 6.4(b) has the crystal structure similar to the crystal structure of famatinite Cu_3SbS_4. In the completely ordered state, the ideal famatinite-type superstructure has two of four simple cubic sublattices constituting the face-centered cubic mixed sublattice filled with one-type atoms. The other two simple cubic sublattices are completed with atoms of two types that are contained in these simple cubic sublattices with the equal concentrations.

There are a lot of types of the superstructures that can be formed by $2A2B1C$ tetrahedral cells in $A_xB_{1-x}C$ alloys with the zinc blende structure. Two of the most simple of them are the short-period superstructures shown in Figures 6.5.

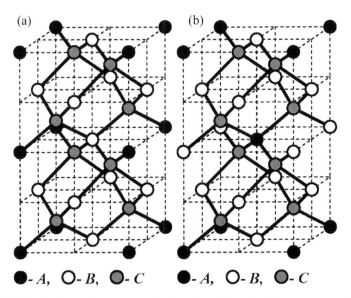

(a) (b)

●- A, ○- B, ◔- C ●- A, ○- B, ◔- C

FIGURE 6.4 The luzonite-type superstructure (a) and famatinite-type superstructure (b).

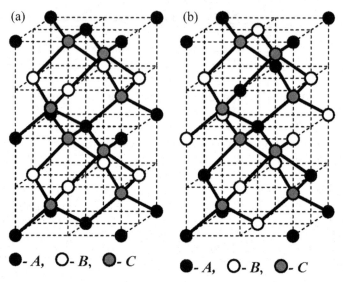

$\bullet$- A, $\bigcirc$- B, ◐- C $\bullet$- A, $\bigcirc$- B, ◐- C

FIGURE 6.5   The CuAu-I superstructure (a) and chalcopyrite-type superstructure (b).

The first type of these superstructures is called the layered tetragonal. So far it is unknown compounds that have such a superstructure. In this case, the mixed sublattice in the completely ordered state consists of two equivalent sublattices formed by alternating (001) layers filled with A and B atoms. The structure of the mixed sublattice is equivalent to the well-known superstructure in $Cu_{0.5}Au_{0.5}$ alloy called the CuAu-I superstructure. Cu and Au crystallize with the face-centered cubic lattices. The equilibrium structure of $Cu_{0.5}Au_{0.5}$ alloy at the temperature below 405 °C is the CuAu-I superstructure. The second type of the superstrucrure is the chalcopyrite-type superstructure (the superstructure has the crystal structure similar to chalcopyrite compounds). In the completely ordered state the mixed sublattice consists of two equivalent sublattices formed by alternating (001) layers filled with A (50%) and B (50%) atoms so that the nearest atoms to A atoms in these layers are atoms B, and vice versa. Besides, a lot of types of more long-period superstructures can be formed by $2A2B1C$ tetrahedral cells. Thus, there are many variants of the long-range order in $A_xB_{1-x}C$ ternary alloys. Moreover, there may be the structures consisting of $2A2B1C$ tetrahedral cells without the long-range order.

In the case of $A_{0.5}B_{0.5}C$ alloys with the wurtzite structure, two types are possible of the long-range order in which the crystal structure is composed of $2A2B1C$ tetrahedral cells. The first type of the long-range order, shown in Figure 6.6, is represented by two sets of alternating $(11\bar{2}0)$ atomic planes filled with A and B atoms, respectively.

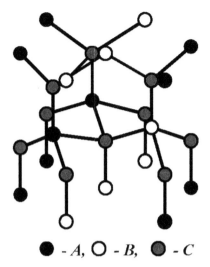

$\bullet$ - A, $\bigcirc$ - B, $\bullet$ - C

FIGURE 6.6 The superstructure with two (0001) oriented atomic layers of the mixed sublattice filled A and B atoms.

The second type of the long-range order is equivalent to the structure of α-BeSiN$_2$. α-BeSiN$_2$ crystallizes in the ordered wurtzite-type structure, which can be described as a C1-type distortion of the idealized filled C9 structure of β-cristobalite [9]. Two (0001) oriented atomic layers of the mixed sublattice with this type of ordering are shown in Figure 6.7(a) and (b).

The superstructures in $A_xB_{1-x}C$ alloys may also be formed by $3A1B1C$ (50%) and $1A3B1C$ (50%) tetrahedral cells. In $A_{0.5}B_{0.5}C$ alloys with the zinc

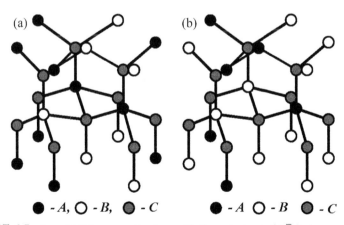

(a) $\bullet$ - A, $\bigcirc$ - B, $\bullet$ - C (b) $\bullet$ - A $\bigcirc$ - B $\bullet$ - C

FIGURE 6.7 (a and b) The superstructure with the ordering in (11$\bar{2}$0) planes.

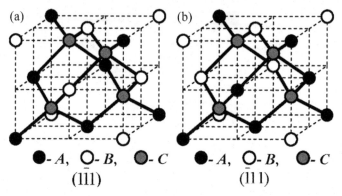

FIGURE 6.8   (a and b) The superstructures of the "CuPt"-type.

blende structure, it is the "CuPt"-type superstructure in the mixed sub-lattice. The disordered $Cu_{0.5}Pt_{0.5}$ alloy with the face-centered cubic structure undergoes during cooling the phase transition to the ordered state with the occurrence of the superstructure. In the completely ordered state, $Cu_{0.5}Pt_{0.5}$ alloy consists of two equivalent sublattices, formed by alternating (111) oriented layers filled with Cu and Pt atoms. The superstructure of the "CuPt"-type with $(1\bar{1}1)$ and $(\bar{1}11)$ oriented layers filled with A and B atoms is shown in Figure 6.8(a) and (b).

The strain energy of the completely ordered "CuPt"-type $A_{0.5}B_{0.5}C$ semiconductor alloys is insignificantly less than the strain energy of the disordered alloys. Therefore, such ordered alloys formed by using the non-equilibrium growth methods such as molecular beam epitaxy or metalorganic chemical vapor deposition are in the metastable or unstable state with respect to the disordered state. In $GaAs_xP_{1-x}$, $In_xGa_{1-x}P$, $In_xGa_{1-x}As$, and $GaSb_xAs_{1-x}$ alloys grown on the (001) substrates, the "CuPt"-type of the superstructure was found only in the $[\bar{1}11]$ or $[1\bar{1}1]$ or both directions among four equivalent <111>B directions: $[\bar{1}11]$, $[1\bar{1}1]$, $[11\bar{1}]$, and $[\bar{1}\bar{1}\bar{1}]$. On the other hand, in $GaAs_xP_{1-x}$, $In_xGa_{1-x}P$, $In_xGa_{1-x}As$, and $GaSb_xAs_{1-x}$, "CuPt"-type of the superstructure does not form if the epitaxial layers were grown on the (111)B or (110) oriented substrates, where (111)B represents four equivalent surfaces: $(\bar{1}11)$, $(1\bar{1}1)$, $(11\bar{1})$, and $(\bar{1}\bar{1}\bar{1})$. The long-range order in the wurtzite $A_{0.5}B_{0.5}C$ alloys formed by $3A1B1C$ (50%) and $1A3B1C$ (50%) tetrahedral cells, shown in Figure 6.9, is represented by two sets of the alternating (0001) atomic layers of the mixed sublattice filled with the different atoms.

The hexagonal closed-packed structure of the completely ordered mixed sublattice is equivalent to the WC superstructure. Therefore it is called the "WC" superstructure.

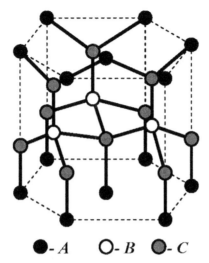

- A   O- B   - C

FIGURE 6.9   The "WC" superstructure.

6.4 ORDER–DISORDER TRANSITIONS

The occurrence of the long-range order, as well as its disappearance, is a result of either a discontinuous or a continuous phase transition. In the case of the discontinuous phase transition, there is a latent heat and there are discontinuities of the enthalpy, the entropy, and the volume (density) at the transition temperature. The differences between the free energies of the ordered and disordered alloys $f^O - f^D$ for alloys undergoing the discontinuous phase transition during cooling of the disordered alloy as functions of the long-range order parameter for various temperatures are shown in Figure 6.10.

At seen from Figure 6.10, the disordered as well as ordered alloy may be in the thermodynamically stable, metastable, or unstable states.

The curve D corresponds to the condition of the stable disordered state and unstable ordered states given by the condition:

$$\frac{\partial f(r_D, T_D)}{\partial r} = 0, \quad \frac{\partial f(r > r_D, T_D)}{\partial r} > 0,$$

where r_D and T_D are the long-range parameter of the disordered alloy and the temperature, correspondingly. The long-range parameter r_D of the disordered alloy is in the origin of coordinates.

The condition of the stable disordered and metastable ordered states with the long-range order r_{MO} at temperature T_{MO} ($T_{MO} < T_D$) is demonstrated by the curve MO and is given by:

$$f(r_D, T_{MO}) < f(r_{MO}, T_{MO}),$$

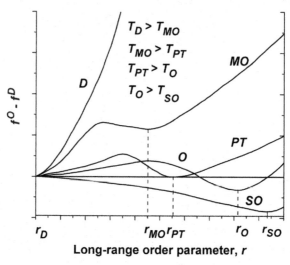

FIGURE 6.10 The differences between the free energies of the ordered and disordered alloys undergoing the discontinuous-phase transition as functions of the long-range order parameter.

$$\frac{\partial f(r_D, T_{MO})}{\partial r} = \frac{\partial f(r_{MO}, T_{MO})}{\partial r} = 0,$$

where r_{MO} and T_{MO} are the long-range parameter of the metastable ordered alloy and the temperature, respectively.

The free energies at temperature T_{MO} of disorderded and ordered states attain the absolute and relative minimums, respectively. The other ordered alloys are in the unstable states with respect to disordering or to the increase or to the decrease of ordering. The ordered alloys with the long-range order parameter from r_D to the parameter determined by the condition:

$$\frac{\partial f(r, T_{MO})}{\partial r} = 0,$$

$$\frac{\partial^2 f(r, T_{MO})}{\partial r^2} < 0$$

are in the unstable state with respect to disordering. The alloys with the long-range order parameter in the range from the parameter defined by the relation:

$$\frac{\partial f(r, T_{MO})}{\partial r} = 0,$$

$$\frac{\partial^2 f(r, T_{MO})}{\partial r^2} < 0$$

to the parameter r_{MO} are in the unstable state with respect to the increase of the long-range order parameter up to $r = r_{MO}$. The alloys with the long-range order parameter $r > r_{MO}$ are in the unstable state with respect to the decrease of ordering up to $r = r_{MO}$.

The phase transition condition between the disordered and ordered states is presented by the curve PT and written as:

$$f(r_D, T_{PT}) = f(r_{PT}, T_{PT}),$$

$$\frac{\partial f(r_D, T_{PT})}{\partial r} = \frac{\partial f(r_{PT}, T_{PT})}{\partial r} = 0,$$

where r_{PT} is the long-range order parameter of the ordered alloy at the phase transition temperature T_{PT}. The ordered alloys with the long-range order parameter $r \neq r_{PT}$ are in the unstable states with respect to disordering or to the increase or to the decrease of ordering. The ordered alloys with the long-range order parameter from r_D to the parameter satisfying the condition:

$$\frac{\partial f(r, T_{PT})}{\partial r} = 0,$$

$$\frac{\partial^2 f(r, T_{PT})}{\partial r^2} < 0$$

are in the unstable state with respect to disordering. The alloys with the long-range order parameter from the parameter described by the condition:

$$\frac{\partial f(r, T_{PT})}{\partial r} = 0,$$

$$\frac{\partial^2 f(r, T_{MO})}{\partial r^2} < 0$$

to the parameter r_{PT} are in the unstable state with respect to the increase of the long-range order parameter up to $r = r_{PT}$. The alloys with the long-range order parameter $r > r_{PT}$ are in the unstable state with respect to the decrease of ordering up to $r = r_{PT}$. The condition of the metastable disordered and stable ordered states corresponding to the curve O is given by:

$$f(r_D, T_O) > f(r_O, T_O),$$

$$\frac{\partial f(r_D, T_O)}{\partial r} = \frac{\partial f(r_O, T_O)}{\partial r} = 0,$$

$$\frac{\partial^2 f(r_D, T_O)}{\partial r^2} > 0,$$

where r_O and T_O are the long-range order parameter and the temperature, correspondingly. The ordered alloys with the long-range order parameter from r_D to the parameter determined by the condition:

$$\frac{\partial f(r, T_O)}{\partial r} = 0,$$

$$\frac{\partial^2 f(r, T_O)}{\partial r^2} < 0$$

are in the unstable state with respect to disordering. The alloys with the long-range order parameter from the parameter given by the condition:

$$\frac{\partial f(r, T_O)}{\partial r} = 0, \quad (r_D < r < r_O)$$

$$\frac{\partial^2 f(r, T_O)}{\partial r^2} < 0$$

to the parameter r_O are in the unstable state with respect to the increase of the long-range order parameter up to $r = r_O$. The alloys with the long-range order parameter $r > r_O$ are in the unstable state with respect to the decrease of ordering up to $r = r_O$.

The condition of the spinodal ordering when the alloy reaches the limit of stability with respect to the formation of the superstructure is:

$$\frac{\partial f(r_D, T_{SO})}{\partial r} = 0,$$

$$\frac{\partial^2 f(r_D, T_{SO})}{\partial r^2} = 0,$$

where T_{SO} is the spinodal ordering temperature. The free energy of the alloy at the spinodal disordering temperature is shown by the curve SO. These temperatures satisfy the following condition $T_D > T_{MO} > T_{PT} > T_O > T_{SO}$ according to the cooling mode.

The curves SO, O, PT, and MO in Figure 6.10 correspond also to the ordered states of the heated ordered alloy. The ordered alloy is in the stable state with respect to disordering at temperatures T_{SO} and T_O. Temperature T_{PT} is the temperature of the order–disorder phase transition of the ordered alloy. At the temperature T_{MO}, the heated ordered alloy is also in the metastable state with respect to disordering. The ordered alloy becomes unstable with respect to disordering at the temperature of the spinodal disordering (the temperature of the stability limit with respect to the disappearance of the superstructure) when the infinitesimal disorder decreases the free energy of the ordered alloy.

The condition of the spinodal disordering is:

$$\frac{\partial f(r_{SD}, T_{SD})}{\partial r} = 0,$$

$$\frac{\partial^2 f(r_{SD}, T_{SD})}{\partial r^2} = 0,$$

where r_{SD} ($r_{SD} < r_{MO}$) and T_{SD} ($T_{SD} > T_{MO}$) are the long-range order parameter of the ordered alloy at the temperature of the spinodal disordering and the temperature of the spinodal disordering, respectively. This condition corresponds to the point of inflection of the free energy as a function of the long-range order.

The formation of the ordered alloy in the metastable or stable state at temperatures higher than the temperature T_{SO} should be a result of finite-size ordering fluctuation occurring. The minimal ordering fluctuation of the long-range order r_{OF} and temperature are obtained by the condition equivalent to the condition of spinodal disordering:

$$\frac{\partial f(r_{OF}, T_{OF})}{\partial r} = 0,$$

$$\frac{\partial^2 f(r_{OF}, T_{OF})}{\partial r^2} = 0,$$

where $r_{OF} = r_{SD}$ and $T_{OF} = T_{SD}$. Such ordering fluctuation in the cooled disordered alloy can continuously develop up to the transformation of the state of the alloy in metastable or stable state. It is well known that the probability of fluctuation increases with its diminution. Moreover, the probability of the same ordering fluctuation is larger at a higher temperature. This happens due to speeding up self-diffusion processes. Thus, in the cooled disordered alloy, the metastable ordered state should initially occur due to the higher probability of the ordering fluctuation. The transformation of the metastable ordered alloy into the stable ordered alloy should take place after the occurrence of the fluctuation determined by the condition:

$$\frac{\partial f(r \geq r_{PT}, T \geq T_{PT})}{\partial r} = 0,$$

$$\frac{\partial^2 f(r \geq r_{PT}, T \geq T_{PT})}{\partial r^2} < 0.$$

The latent heat of the phase transition as well as changes of the internal energy, the entropy, and volume are absent in the case of a continuous order–disorder phase transition. Therefore, the disordered and ordered states may be only in the thermodynamically stable or unstable states with respect to ordering and disordering, respectively. In other words, the alloy reaches the stability limit with respect to the change of its state at the phase transition temperature. The differences between the free energies of the ordered and disordered alloys undergoing the continuous phase transition as a function of the long-range order parameter for various temperatures are shown in Figure 6.11.

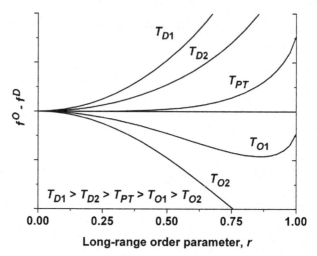

FIGURE 6.11   The differences between the free energies of the ordered and disordered alloys undergoing the continuous-phase transition as a function of the long-range order parameter.

The curves T_{D1} and T_{D2} correspond to the condition of the stable disordered and unstable ordered alloys written as:

$$\frac{\partial f(r_D, T_D)}{\partial r} = 0,$$

$$\frac{\partial f(r > r_D, T_D)}{\partial r} > 0.$$

Under the condition:

$$\frac{df(r_D, T_{PT})}{dr} = 0,$$

$$\frac{d^2 f(r_D, T_{PT})}{dr^2} = 0$$

the alloy undergoes the continuous phase transition between ordered and disordered states and the difference between the free energies at the phase transition temperature is shown by the curve T_{PT}. The difference between the free energies corresponding to the condition of the unstable disordered and stable ordered alloys is demonstrated by the curves T_{O1} and T_{O2} obtained as:

$$f(r_D, T_O) > f(r_O, T_O),$$

$$\frac{\partial f(r_D, T_O)}{\partial r} = \frac{\partial f(r_O, T_O)}{\partial r} = 0,$$

$$\frac{d^2 f(r_D, T_O)}{dr^2} < 0.$$

6.5 DISCONTINUOUS ORDER–DISORDER TRANSITION

The order–disorder transition in $A_x B_{1-x} C$ alloy with the preferential formation of $3A1B1C$ tetrahedral cells is considered. The mixed sublattice of such ternary alloy with the zinc blende structure has the face-centered cubic lattice represented as consisting of two non-equivalent sublattices a and b in which there are $0.75N = N_a$ and $0.25N = N_b$ lattice sites; correspondingly, $N = N_a + N_b$ is the total number of the lattice sites in the mixed sublattice of this alloy. The sublattice a consists of three simple cubic sublattices and sublattice b has a simple cubic structure. In $A_x B_{1-x} C$ completely ordered alloy, the sublattice b is filled with atoms B if $x \leq 0.75$ or contains all atoms B if $x > 0.75$. The ordering degree is characterized by the long-range order parameter r, which is the ratio between the number of atoms B on the sites of the sublattice b (b sites) and the number of the sites of the sublattice b: $r = \frac{N_{B(b)}}{N_b}$. It is assumed that the portion of atoms B on the sites of the sublattice b in the ordered alloy is not less than that in the disordered alloy, and thus $\frac{N_B}{N_a + N_b} \leq r \leq \frac{N_{B(b)}}{N_b}$, where N_B is the number of atoms B. Accordingly, the numbers and concentrations of atoms on the a and b sublattice sites are given, respectively, by:

$$N_{A(a)} = xN - N_{A(b)} = (x - 0.25 + 0.25r)N,$$

$$N_{A(b)} = N_b - N_{B(b)} = 0.25(1 - r)N,$$

$$N_{B(a)} = (1 - x - 0.25r)N,$$

$$N_{B(b)} = 0.25rN,$$

$$x_{A(a)} = \frac{N_{A(a)}}{N_a} = \frac{4x - 1 + r}{3},$$

$$x_{A(b)} = \frac{N_{A(b)}}{N_b} = 1 - r,$$

$$x_{B(a)} = \frac{N_{B(a)}}{N_a} = \frac{4 - 4x - r}{3},$$

$$x_{B(b)} = \frac{N_{B(b)}}{N_b} = r.$$

It is assumed also that the mixed sublattice is undistorted in both the disordered and ordered states. Moreover, it is supposed too that the deformation energy of the angles between bonds belonging to the neighboring tetrahedral cells does not vary as the result of ordering. Accordingly, this energy is not taken into account. There are $4A1C$, $3A1B1C$, $2A2B1C$, $1A3B1C$, and $4B1C$ tetrahedral cells in the alloy considered.

The configurations of tetrahedral cells on the lattice sites are obtained as follows.

$$4A1C: A(a), A(a), A(a), A(b);$$

$$3A1B1C(1): A(a), A(a), A(a), B(b);$$

$$3A1B1C(2): B(a), A(a), A(a), A(b); A(a), B(a), A(a), A(b);$$

$$A(a), A(a), B(a), A(b);$$

$$2A2B1C(1): A(a), A(a), B(a), B(b); A(a), B(a), A(a), B(b);$$

$$B(a), A(a), A(a), B(b);$$

$$2A2B1C(2): B(a), B(a), A(a), A(b); A(a), B(a), B(a), A(b);$$

$$B(a), A(a), B(a), A(b);$$

$$1A3B1C(1): A(a), B(a), B(a), B(b); B(a), A(a), B(a), B(b);$$

$$B(a), B(a), A(a), B(b);$$

$$1A3B1C(2): B(a), B(a), B(a), A(b);$$

$$4B1C: B(a), B(a), B(a), B(b).$$

The internal energy of mixing of the alloy is the strain energy. It is derived from the deformation energies of tetrahedral cells $4A1C, 3A1B1C, 2A2B1C,$ $1A3B1C,$ and $4B1C$. These energies are calculated by using the valence force field model. It is assumed that the structure of the mixed sublattice is undistorted in both the disordered and ordered states. The concentrations of tetrahedral cells with one disposition on the lattice sites are:

$$x_{4A1C} = \left(\frac{4x - 1 + r}{3}\right)^3 (1 - r),$$

$$x_{3A1B1C(1)} = \left(\frac{4x - 1 + r}{3}\right)^3 r,$$

$$x_{3A1B1C(2)} = \frac{4 - 4x - r}{3}\left(\frac{4x - 1 + r}{3}\right)^2 (1 - r),$$

$$x_{2A2B1C(1)} = \left(\frac{4x - 1 + r}{3}\right)^2 \left(\frac{4 - 4x - r}{3}\right) r,$$

$$x_{2A2B1C(2)} = \frac{4x - 1 + r}{3}\left(\frac{4 - 4x - r}{3}\right)^2 (1 - r),$$

$$x_{1A3B1C(1)} = \frac{4x - 1 + r}{3}\left(\frac{4 - 4x - r}{3}\right)^2 r,$$

$$x_{1A3B1C(2)} = \left(\frac{4 - 4x - r}{3}\right)^3 (1 - r),$$

$$x_{4B1C} = \left(\frac{4 - 4x - r}{3}\right)^3 r.$$

The molar internal energy of mixing is:

$$u^M = u_{4A1C} x_{4A1C} + u_{3A1B1C} x_{3A1B1C(1)} + 3 u_{3A1B1C} x_{3A1B1C(2)}$$

$$+ 3 u_{2A2B1C} x_{2A2B1C(1)} + 3 u_{2A2B1C} x_{2A2B1C(2)} + 3 u_{1A3B1C} x_{1A3B1C(1)}$$

$$+ u_{1A3B1C} x_{1A3B1C(2)} + u_{4B1C} x_{4B1C}.$$

The configurational entropy is obtained as a function of the number of configurations. The number of configurations is represented as a product of two factors. The first factor is the number of configurations in the sublattice a, and the second factor is the number of configurations in the sublattice b. This product is:

$$\Omega = \Omega_1 \Omega_2 = \frac{N_a!}{N_{A(a)}! N_{B(a)}!} \frac{N_b!}{N_{A(b)}! N_{B(b)}!}$$

$$= \frac{\left(\frac{3N}{4}\right)!}{\left[(x - \frac{1-r}{4})N\right]! \left[(1 - x - \frac{r}{4})N\right]!} \frac{\left(\frac{N}{4}\right)!}{\left[\frac{(1-r)N}{4}\right]! \left(\frac{rN}{4}\right)!}.$$

The entropy of mixing obtained by using Stirling's formula is given by:

$$s^M = k_B \ln \Omega$$

$$= -R \frac{4x - 1 + r}{4} \ln \frac{4x - 1 + r}{3} - R \frac{4 - 4x - r}{4} \ln \frac{4 - 4x - r}{3}$$

$$- R \frac{1 - r}{4} \ln(1 - r) - R \frac{r}{4} \ln r.$$

The free energy of mixing $f^M = u^M - Ts^M$ demonstrates that semiconductor alloys should be in the discontinuous order–disorder phase transition. The discontinuous phase transition temperature T_{PT} and the long-range order parameter r_{PT} at this temperature satisfy the system of equations:

$$f^M(r_{PT}, T_{PT}) - f^M(r_D, T_{PT}) = 0,$$

$$\frac{\partial f^M(r_{PT}, T_{PT})}{\partial r} = 0.$$

The latent heat of the phase transition is written as:

$$h_{PT} = u^M(r_D) - u^M(r_{PT}).$$

The free energies of mixing of ordered alloys at the transition temperature satisfy the condition:

$$f^M(r_D < r < r_{PT}, T_{PT}) > f^M(r_{PT}, T_{PT}),$$

$$f^M(r > r_{PT}, T_{PT}) > f^M(r_{PT}, T_{PT}).$$

Spinodal ordering occurs in a disordered alloy when it reaches the limit of thermodynamic stability with respect to the formation of the superstructure. In other words, it happens when the occurrence of the negligibly small ordering fluctuation decreases the free energy of mixing. The condition of this limit of the thermodynamic stability is:

$$\delta f^M = f^M(r_D + \delta r, T_{SO}) - f^M(r_D, T_{SO}) = 0,$$

where T_{SO} is the spinodal ordering temperature. Accordingly, the condition determining the spinodal ordering temperature obtained by using the Taylor's series expansion is:

$$\frac{d^2 f^M(r_D, T_{SO})}{dr^2} = 0, \quad \text{since} \quad \frac{df^M(r_D, T)}{dr} = 0.$$

Spinodal disordering occurs in the ordered alloy when the alloy reaches the limit of the thermodynamic stability with respect to vanishing of the superstructure. In other words, spinodal disordering occurs when the negligibly small disordering fluctuation decreases the free energy of mixing. The condition of this limit of the stability is given by:

$$\delta f^M = f^M(r_{SD} - \delta r, T_{SD}) - f^M(r_{SD}, T_{SD}) = 0,$$

where r_{SD} is the long-range order parameter at the spinodal disordering temperature. Accordingly, the condition determining the spinodal disordering temperature and the parameters of the long-range order at this temperature obtained by using the Taylor's series expansion is given by the system of equations:

$$\frac{df^M(r_{SD}, T_{SD})}{dr} = 0,$$

$$\frac{d^2 f^M(r_{SD}, T_{SD})}{dr^2} = 0.$$

As an example, the characteristics of the discontinuous phase transition are estimated here for $GaSb_{0.75}As_{0.25}$ alloy. The strain energies of the tetrahedral cells in this alloy are $u_{4Sb1Ga} = 2.931$ kJ mol^{-1}, $u_{3Sb1As1Ga} = 2.618$ kJ mol^{-1}, $u_{2Sb2As1Ga} = 7.571$ kJ mol^{-1}, $u_{1Sb3As1Ga} = 17.456$ kJ mol^{-1}, and $u_{4As1Ga} = 32.760$ kJ mol^{-1}.

The phase transition temperature, the long-range order parameter, and latent heat of the phase transition of $GaSb_{0.75}As_{0.25}$ alloy are equal, respectively, to $T_{PT} = 237.4\,°C$, $r_{PT} = 0.611$, and $h_{PT} = 450$ J mol^{-1}. The spinodal ordering temperature is $T_{SO} = 191.5\,°C$. The spinodal disordering temperature and the long-range order parameter at this temperature are equal, respectively, to $T_{SD} = 243\,°C$ and $r_{SD} = 0.519$ for $GaSb_{0.75}As_{0.25}$.

6.6 CONTINUOUS ORDER–DISORDER TRANSITION

The order–disorder transition in $A_xB_{1-x}C$ ternary alloy with the preferential formation of $2A2B1C$ tetrahedral cells in the ordered state is considered. The ordering conditions are calculated by minimizing the free energy of zinc blende or wurtzite $A_xB_{1-x}C$ alloy. The redistribution of A and B atoms changes only the free energy of mixing, so it is only taken into account in the calculations. The free energy of mixing ($f^M = u^M - Ts^M$) is a sum of the strain energy caused by the difference in the lattice parameters of AC and BC and the configurational entropy term. The crystal structure is represented as a set of the tetrahedral cells of the same size consisting of four cations in their vertices and anion atoms in the centers. There are $4A1C$, $3A1B1C$, $2A2B1C$, $1A3B1C$, and $4B1C$ tetrahedral cells. It is assumed that the mixed sublattice is undistorted in both the disordered and ordered states. Besides, it is supposed that the deformation energy of the angles between bonds belonging to the neighboring tetrahedral cells does not vary as a result of ordering. Accordingly, this energy is not considered.

To introduce the ordering into the model, the mixed sublattice is represented as two crystallographically equivalent sublattices a and b. In $A_xB_{1-x}C$ disordered alloy, each atom in the mixed sublattice regardless of the type (A or B) may be placed in either of two sublattices with equal probability. In the case of complete ordering (the ideal superstructure), all minority atoms belong to one sublattice (i.e., in the ideal $A_xB_{1-x}C$ superstructure with $x < 0.5$ all A atoms are positioned on the sites of the sublattice a, whereas in the case of $x > 0.5$ all B atoms belong to the sublattice b). In the intermediate case, the ordering is represented by the long-range order parameter r, $(0 < r < 1)$, with $r = 0$ corresponding to the disordered alloy and $r = 1$ to the ideal superstructure. As usual, the occupations of a site in the sublattice a by the atom A or of a site in the sublattice b by the atom B are referred to as the "right" occupations, the others being the "wrong" occupations. The probabilities of the "right" occupations in the ordered alloy are greater than in the disordered alloy. The arrangement of atoms in each of the sublattices a and b is assumed to be random. The atomic concentrations in the sublattices as functions of the composition and the long-range order parameter are:

$$x_{A(a)} = x + yr, \quad x_{A(b)} = x - yr,$$

$$x_{B(a)} = 1 - x - yr, \quad x_{B(b)} = 1 - x + yr,$$

where $y = x$ if $0 \leq x \leq 0.5$ and $y = 1 - x$ if $0.5 \leq x \leq 1$. The configurations of tetrahedral cells over the lattice sites are given by:

$$4A1C: A(a), A(a), A(b), A(b);$$

$$3A1B1C(1): A(a), A(a), A(b), B(b); A(a), A(a), B(b), A(b);$$

$$3A1B1C(2): B(a), A(a), A(b), A(b); A(a), B(a), A(a), A(b);$$

$$2A2B1C(1): A(a), A(a), B(b), B(b);$$

$$2A2B1C(2): A(a), B(a), A(b), B(b); A(a), B(a),$$

$$B(b), A(b); B(a), A(a), A(b), B(b);$$

$$B(a), A(a), B(b), A(b);$$

$$2A2B1C(3): B(a), B(a), A(b), A(b);$$

$$1A3B1C(1): A(a), B(a), B(b), B(b); B(a), A(a), B(b), B(b);$$

$$1A3B1C(2): B(a), B(a), A(b), B(b); B(a), B(a), B(b), A(b);$$

$$4B1C: B(a), B(a), B(b), B(b).$$

The concentrations of tetrahedral cells with one disposition over the lattice sites are:

$$x_{4A1C} = \left(x^2 - y^2 r^2\right)^2,$$

$$x_{3A1B1C(1)} = (x + yr)^2 (x - yr)(1 - x + yr),$$

$$x_{3A1B1C(2)} = (1 - x - yr)(x + yr)(x - yr)^2,$$

$$x_{2A2B1C(1)} = (x + yr)^2 (1 - x + yr)^2,$$

$$x_{2A2B1C(2)} = \left(x^2 - y^2 r^2\right)\left[(1 - x)^2 - y^2 r^2\right],$$

$$x_{2A2B1C(3)} = (1 - x - yr)^2 (x - yr)^2,$$

$$x_{1A3B1C(1)} = (x + yr)(1 - x - yr)(1 - x + yr)^2,$$

$$x_{1A3B1C(2)} = (1 - x - yr)^2 (x - yr)(1 - x + yr),$$

$$x_{4B1C} = (1 - x - yr)^2 (1 - x + yr)^2,$$

where $x_{3A1B1C(1)}$ is the concentration of $3A1B1C(1)$ tetrahedral cells. The internal energy of mixing is:

$$u^M = u_{4A1C} x_{4A1C} + 2u_{3A1B1C}\left(x_{3A1B1C(1)} + x_{3A1B1C(2)}\right)$$
$$+ u_{2A2B1C}\left(x_{2A2B1C(1)} + 4x_{2A2B1C(2)} + x_{2A2B1C(3)}\right)$$
$$+ 2u_{1A3B1C}\left(x_{1A3B1C(1)} + x_{1A3B1C(2)}\right) + u_{4B1C} x_{4B1C}$$

where u_{4A1C} is the molar strain energy of $4A1C$ tetrahedral cell.

The entropy of mixing is represented by the number of the atomic permutations:

$$\Omega = \frac{\left(\frac{1}{2}N_{Av}\right)!}{\left[\frac{1}{2}N_{Av}(x+yr)\right]!\left[\frac{1}{2}N_{Av}(1-x-yr)\right]!}$$

$$\times \frac{\left(\frac{1}{2}N_{Av}\right)!}{\left[\frac{1}{2}N_{Av}(x-yr)\right]!\left[\frac{1}{2}N_{Av}(1-x+yr)\right]!},$$

where N_{Av} is Avogadro's number (the total number of A and B atoms). Accordingly, the entropy of mixing is:

$$s^M = k_B \ln \Omega$$

$$= -\frac{1}{2}R(x+yr)\ln(x+yr) - \frac{1}{2}R(1-x-yr)\ln(1-x-yr)$$

$$-\frac{1}{2}R(x-yr)\ln(x-yr) - \frac{1}{2}R(1-x+yr)\ln(1-x+yr).$$

The condition for the free energy given by $\frac{df^M(r_D,T)}{dr} = 0$ is fulfilled at any temperature. Therefore, the order–disorder transition temperature is calculated from the condition:

$$\frac{d^2 f^M(r_D, T_{PT})}{dr^2} = 0.$$

Thus, the order–disorder transition is the continuous-phase transition without the latent heat of transition. The order–disorder transition temperature is the temperature of the transformation of the stable state to the unstable state with respect to the occurrence as well as vanishing of the superstructure. For example, the order–disorder transition temperature in $GaSb_{0.5}As_{0.5}$ is equal to 283 °C. The long-range order parameter of the ordered alloy is obtained by the condition:

$$\frac{df^M(r > r_D, T < T_{PT})}{dr} = 0.$$

6.7 CARBON AND Sn IN Ge

Carbon and Sn are the isoelectronic substitutional impurities in Ge, since the majority of them are over the lattice sites of the Ge-rich matrix of Ge doped with carbon or Sn. Therefore Ge doped with carbon and Sn is considered a Ge-rich $C_x Sn_y Ge_{1-x-y}$ ternary alloy. Small carbon and large Sn atoms as compared with Ge atoms dilate and shrink, respectively, the crystal lattice of $C_x Sn_y Ge_{1-x-y}$ alloy. It is reasonable to expect that the formation of clusters containing carbon and Sn atoms in Ge doped with carbon and Sn may compensate the dilation and shrinking caused by the isolated impurities. Further, the consideration of the distortions of the crystal lattice, the internal strain energy, and the compensation of the internal strains after the formation of the clusters is fulfilled for the alloys with the concentrations of carbon and Sn in the dilute limit (x, $y \ll 1$). Neither carbon nor Sn atoms form chemical bonds with Ge. Hence, the Helmholtz free energy of $C_xSn_yGe_{1-x-y}$ alloy is:

$$f = f_C^0 + f_{Sn}^0 + f_{Ge}^0 + u^{Int.Str.} - Ts^{Conf.},$$

where f_C^0, $u^{Int.Str.}$ and $s^{Conf.}$ are the free energy of carbon atoms, the internal strain energy, and the configurational entropy, correspondingly. The occurrence of carbon- and Sn-containing clusters may change only the internal strain energy and the configurational entropy. Isolated carbon and Sn atoms form around themselves the quadruples of 1C3Ge1Ge and 1Sn3Ge1Ge tetrahedral cells, respectively, with Ge atoms in their centers in the diamond structure of the Ge-rich matrix. These quadruples are shown in Figures 6.12 and 6.13.

The distortions of the crystal lattice and the internal strain energy due to isolated carbon and Sn atoms are obtained by the way the internal strain energy is represented as a sum of three energies. The deformation energies of 1C3Ge1Ge and 1Sn3Ge1Ge tetrahedral cells are the first terms in the sums. Due to the symmetry of the diamond structure, 1C3Ge1Ge and 1Sn3Ge1Ge cells in the quadruples are geometrically undistorted tetrahedrons of the same size. The deformation energies of the bond and the angle between bonds in the tetrahedral cell obtained by using the valence force field model (see Section 6.2) are given, respectively, by:

$$u_b = \frac{3\alpha\left(R^2 - r^2\right)^2}{8R^2}, \tag{6.7.1}$$

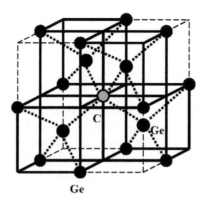

Ge

FIGURE 6.12　Quadruple of 1C3Ge1Ge tetrahedral cells in the Ge-rich matrix.

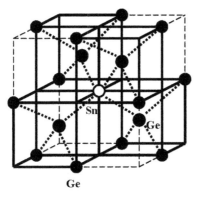

FIGURE 6.13   Quadruple of 1Sn3Ge1Ge tetrahedral cells in the Ge-rich matrix.

$$u_a = \frac{3\beta \left(R^2 \cos \varphi_0 - r_1 r_2 \cos \varphi \right)^2}{8R^2}, \qquad (6.7.2)$$

where R is the distance between the nearest neighbors in the undistorted crystal; r, r_1, and r_2 are the distances between the nearest neighbors in the distorted crystal, and $\varphi_0 = 109.47^\circ$ ($\cos \varphi_0 = -1/3$) and φ are the angles between the bonds in the undistorted and distorted crystals, respectively. One tetrahedral cell contains four bonds and six angles between bonds. The deformation energy $u_{1(C)}$ of the quadruple of 1C3Ge1Ge tetrahedral cells around a carbon atom is estimated as follows. The carbon atom is at the origin of the coordinates. The central Ge(C) atom in the tetrahedral cell shown in Figure 6.14 is at coordinates $x = y = z = l$, where (C) denotes that the tetrahedral cell is formed by the isolated carbon atom.

The size of the edge of the cube of the tetrahedral cell is equal to d, and there are three of the same GeGe(C) bonds and one CGe(C) bond in this cell. There are three of the same CGeGe(C) and three of the same

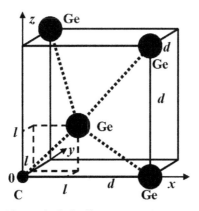

FIGURE 6.14   1C3Ge1Ge tetrahedral cell.

GeGeGe(C) angles between the bonds in the 1C3Ge1Ge tetrahedral cell. The lengths of the bonds and cosines of the angles between bonds are, respectively:

$$r_{CGe(C)} = \sqrt{3}l, \quad r_{GeGe(C)} = \sqrt{3l^2 - 4dl + 2d^2},$$

$$\cos \varphi_{CGeGe(C)} = \frac{3l - 2d}{\sqrt{3}\sqrt{l^2 + 2(d - l)^2}},$$

$$\cos \varphi_{GeGeGe(C)} = \frac{-2l(d - l) + (d - l)^2}{\sqrt{2(d - l)^2 + l^2}\sqrt{2(d - l)^2 + l^2}}.$$

The angles between bonds are obtained by using the formula for the cosine of the angle α between the vectors $\mathbf{a} = (x_1, y_1, z_1)$ and $\mathbf{b} = (x_2, y_2, z_2)$, known from analytic geometry as:

$$\cos \alpha = \frac{x_1 x_2 + y_1 y_2 + z_1 z_2}{\sqrt{x_1^2 + y_1^2 + z_1^2} \times \sqrt{x_2^2 + y_2^2 + z_2^2}}. \tag{6.7.3}$$

The deformation energies of the bonds and the angles between bonds are, respectively:

$$u_{CGe(C)} = \frac{3(\alpha_C + \alpha_{Ge})}{4(R_C + R_{Ge})^2} \left[\left(\frac{R_C + R_{Ge}}{2} \right)^2 - 3l^2 \right]^2,$$

$$u_{GeGe(C)} = \frac{3\alpha_{Ge}}{8R_{Ge}^2} (R_{Ge}^2 - 3l^2 + 4dl - 2d^2)^2,$$

$$u_{CGeGe(C)} = \frac{3(\beta_C + 3\beta_{Ge})}{16(R_C + R_{Ge})R_{Ge}}$$

$$\times \left[-\frac{R_C + R_{Ge}}{6} R_{Ge} - l\sqrt{3l^2 - 4dl + 2d^2} \frac{3l - 2d}{\sqrt{l^2 + 2(d - l)^2}} \right]^2,$$

$$u_{GeGeGe(C)} = \frac{3\beta_{Ge}}{8R_{Ge}^2} \left[-\frac{1}{3}R_{Ge}^2 - (3l^2 - 4dl + 2d^2)\frac{3l^2 - 4dl + d^2}{l^2 + 2(d - l)^2} \right]^2,$$

where it is supposed that the bond-stretching elastic constant of CGe(C) bond is equal to $\frac{\alpha_C + \alpha_{Ge}}{2}$, the length of the undistorted CGe(C) bond is $\frac{R_C + R_{Ge}}{2}$, and the bond-bending elastic constant of CGeGe(C) angle is $\frac{\beta_C + 3\beta_{Ge}}{4}$. Finally, the deformation energy of 1C3Ge1Ge tetrahedral cell $u_{1(C)}$ is:

$$u_{1(C)} = u_{CGe(C)} + 3u_{GeGe(C)} + 3u_{CGeGe(C)} + 3u_{GeGeGe(C)}.$$

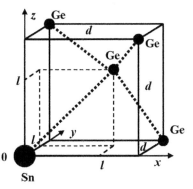

FIGURE 6.15 1Sn3Ge1Ge tetrahedral cell.

The distortions of the crystal lattice and the internal strain energy of the 1Sn3Ge1Ge tetrahedral cell formed by the isolated Sn atom shown in Figure 6.15 is estimated in a similar way. The Sn atom is at the origin of the coordinates.

The central Ge(Sn) atom in the cell is at coordinates $x = y = z = l$. The size of the edge of the cube with the tetrahedral cell is equal to d, and there are three of the same GeGe(Sn) bonds and one SnGe(Sn) bond in this cell. There are three of the same SnGeGe(Sn) and three of the same GeGeGe(Sn) angles between the bonds. The lengths of the bonds and the cosines of the angles between the bonds are, respectively:

$$r_{SnGe(Sn)} = \sqrt{3}l,$$

$$r_{GeGe(Sn)} = \sqrt{3l^2 - 4dl + 2d^2},$$

$$\cos \varphi_{SnGeGe(Sn)} = \frac{3l - 2d}{\sqrt{3}\sqrt{l^2 + 2(d - l)^2}},$$

$$\cos \varphi_{GeGeGe(Sn)} = \frac{-2l(d - l) + (d - l)^2}{\sqrt{2(d - l)^2 + l^2}\sqrt{2(d - l)^2 + l^2}},$$

The deformation energies of the bonds and the angles between the bonds are, respectively:

$$u_{SnGe(Sn)} = \frac{3(\alpha_{Sn} + \alpha_{Ge})}{4(R_{Sn} + R_{Ge})^2}\left[\left(\frac{R_{Sn} + R_{Ge}}{2}\right)^2 - 3l^2\right]^2,$$

$$u_{GeGe(Sn)} = \frac{3\alpha_{Ge}}{8R_{Ge}^2}\left(R_{Ge}^2 - 3l^2 + 4dl - 2d^2\right)^2,$$

$$u_{SnGeGe(Sn)} = \frac{3(\beta_{Sn} + 3\beta_{Ge})}{16(R_{Sn} + R_{Ge})R_{Ge}}$$

$$\times \left[-\frac{R_{Sn} + R_{Ge}}{6} R_{Ge} - l\sqrt{3l^2 - 4dl + 2d^2} \frac{3l - 2d}{\sqrt{l^2 + 2(d-l)^2}} \right]^2,$$

$$u_{GeGeGe(Sn)} = \frac{3\beta_{Ge}}{8R_{Ge}^2} \left[-\frac{1}{3}R_{Ge}^2 - (3l^2 - 4dl + 2d^2) \frac{3l^2 - 4dl + d^2}{l^2 + 2(d-l)^2} \right]^2,$$

where it is supposed that the bond-stretching elastic constant of the SnGe(Sn) bond is $\frac{\alpha_{Sn} + \alpha_{Ge}}{2}$, the length of the undistorted SnGe(Sn) bond is $\frac{R_{Sn} + R_{Ge}}{2}$, and the bond-bending elastic constant of the SnGeGe(Sn) angle is $\frac{\beta_{Sn} + 3\beta_{Ge}}{4}$. The deformation energy of the 1Sn3Ge1Ge tetrahedral cell is:

$$u_{1(Sn)} = u_{SnGe(Sn)} + 3u_{GeGe(Sn)} + 3u_{SnGeGe(Sn)} + 3u_{GeGeGe(Sn)}.$$

The sums of the deformation energies of six angles belonging to the different 1C3Ge1Ge and 1Sn3Ge1Ge tetrahedral cells in the quadruples of the tetrahedral cells are the second items of the sums, representing the internal strain energies caused by isolated carbon and Sn atoms, respectively. The angles between bonds belonging to the different tetrahedral cells are undistorted, as seen in Figures 6.12 and 6.13. However, the bonds are distorted. Hence, there are the deformation energies of such angles. The deformation energies $u_{2(C)}$ and $u_{2(Sn)}$ of the angles GeCGe(C) and GeSnGe(Sn) belonging to the different 1C3Ge1Ge and 1Sn3Ge1Ge tetrahedral cells are, respectively:

$$u_{2(C)} = \frac{\beta_C + \beta_{Ge}}{12(R_C + R_{Ge})^2} \left[\left(\frac{R_C + R_{Ge}}{2} \right)^2 - 3l^2 \right]^2,$$

$$u_{2(Sn)} = \frac{\beta_{Sn} + \beta_{Ge}}{12(R_{Sn} + R_{Ge})^2} \left[\left(\frac{R_{Sn} + R_{Ge}}{2} \right)^2 - 3l^2 \right]^2$$

where it is also supposed that the bond-stretching elastic constants between the different atoms and bond-bending elastic constants between the different bonds are equal to the average values, and the length of the undistorted SnGe(Sn) bond is $\frac{R_{Sn} + R_{Ge}}{2}$.

The internal strain energies of the crystal lattices outside the quadruples of the tetrahedral cells are the third item of the sums representing the strain energies due to the isolated carbon and Sn atoms. This third item is considered as the strain energy of the crystal lattice outside the sphere, centered at the carbon or Sn atom due to the symmetry of the quadruples. The radius of the sphere is the diagonal of the face of the cube with the tetrahedron cell equal to $\sqrt{2}d$ (Figures 6.12 and 6.13), with the origin of coordinates at the carbon or Sn atom. The radial displacement of the atoms of the crystal lattice outside the quadruple at the distance r is assumed to be expressed as $u \propto r^{-2}$, similarly to the displacement in the elastic medium [10]. The projections of the vector $\mathbf{r}$ on the coordinate axes are denoted by r_x, r_y, and r_z. If the displacement u of $\sqrt{2}d$ is:

$$u\left(\sqrt{2}d\right) = \sqrt{2}\left(d - \frac{2}{\sqrt{3}}R_{Ge}\right),$$

where R_{Ge} is the distance between the nearest neighbors in Ge, then the displacement with distance r is:

$$u(r) = u\left(\sqrt{2}d\right)\frac{2d^2}{r^2}.$$

The elastic energy density of the deformed crystal with the cubic structure is:

$$\begin{aligned}
\varepsilon &= \frac{1}{2}C_{11}\left(e_{xx}^2 + e_{yy}^2 + e_{zz}^2\right) + C_{12}(e_{xx}e_{yy} + e_{yy}e_{zz} + e_{zz}e_{xx}) \\
&+ \frac{1}{2}C_{44}\left(e_{xy}^2 + e_{yz}^2 + e_{zx}^2\right) = \frac{1}{2}C_{11}(e_{xx} + e_{yy} + e_{zz})^2 \\
&- (C_{11} - C_{12})(e_{xx}e_{yy} + e_{yy}e_{zz} + e_{zz}e_{xx}) + \frac{1}{2}C_{44}\left(e_{xy}^2 + e_{yz}^2 + e_{zx}^2\right),
\end{aligned}$$

(6.7.4)

where C_{11} is the stiffness coefficient of the crystal and e_{ij} $(i, j = x, y, z)$ are the components of the strain tensor. The components of the strain tensor of the radially strained elastic medium in the spherical coordinate system are:

$$e_{xx} = \frac{\partial u_x}{\partial x} = 2u\left(\sqrt{2}d\right)d^2\frac{r^2 - 3r_x^2}{r^5}, \tag{6.7.5}$$

$$e_{yy} = \frac{\partial u_y}{\partial y} = u\left(\sqrt{2}d\right)d^2\frac{r^2 - 3r_y^2}{r^5}, \tag{6.7.6}$$

$$e_{zz} = \frac{\partial u_z}{\partial z} = u\left(\sqrt{2}d\right)d^2\frac{r^2 - 3r_z^2}{r^5}, \tag{6.7.7}$$

$$e_{xy} = \frac{\partial u_x}{\partial y} + \frac{\partial u_y}{\partial x} = -12u\left(\sqrt{2}d\right)d^2\frac{r_x r_y}{r^5}, \tag{6.7.8}$$

$$e_{yz} = \frac{\partial u_y}{\partial z} + \frac{\partial u_z}{\partial y} = -12u\left(\sqrt{2}d\right)d^2\frac{r_y r_z}{r^5}, \tag{6.7.9}$$

$$e_{zx} = \frac{\partial u_z}{\partial x} + \frac{\partial u_x}{\partial z} = -12u\left(\sqrt{2}d\right)d^2\frac{r_z r_x}{r^5}, \tag{6.7.10}$$

where $r_x = r\sin\theta\cos\varphi$, $r_y = r\sin\theta\sin\varphi$, $r_z = r\cos\theta$.

The internal strain energy of Ge-rich $C_xSn_yGe_{1-x-y}$ alloy outside the quadruple is:

$$u_3 = \int_{\sqrt{2}d}^{\infty} \varepsilon(r)d^3r$$

$$= -4(C_{11} - C_{12})d^4u\left(\sqrt{2}d\right)^2$$

$$\times \iiint \left(\frac{1 - 3\sin^2\theta\cos^2\varphi}{r^2} \times \frac{1 - 3\sin^2\theta\sin^2\varphi}{r^2}\right)\sin^5\theta \, dr d\theta d\varphi$$

$$- 4(C_{11} - C_{12})d^4u\left(\sqrt{2}d\right)^2 \iiint \left(\frac{1 - 3\sin^2\theta\sin^2\varphi}{r^2} \frac{1 - 3\cos^2\theta}{r^2}\right)$$

$$\times \sin\theta \, dr d\theta d\varphi - 4(C_{11} - C_{12})d^4u\left(\sqrt{2}d\right)^2$$

$$\times \iiint \left(\frac{1 - 3\sin^2\theta\cos^2\varphi}{r^2}\frac{1 - 3\cos^2\theta}{r^2}\right)\sin\theta \, dr d\theta d\varphi$$

$$- 2C_{44}d^4u\left(\sqrt{2}d\right)^2 \iiint \left(\frac{\sin^4\theta\cos^2\varphi\sin^2\varphi}{r^4} + \frac{\sin^2\theta\cos^2\theta}{r^4}\right)$$

$$\times \sin\theta \, dr d\theta d\varphi - 2C_{44}d^4u\left(\sqrt{2}d\right)^2$$

$$\times \iiint \left(\frac{\sin^4\theta\cos^2\varphi\sin^2\varphi}{r^4} + \frac{\sin^2\theta\cos^2\theta}{r^4}\right)\sin\theta \, dr d\theta d\varphi$$

$$= \frac{16\sqrt{2}}{5}\pi(C_{11} - C_{12} + 3C_{44})d\left(d - \frac{2}{\sqrt{3}}R_{Ge}\right)^2,$$

$$(6.7.11)$$

where C_{11} is the stiffness coefficient of Ge.

The internal strain energies due to isolated carbon and Sn atoms in Ge are:

$$u_C = 4u1_{(C)} + 6u2_{(C)} + u3_{(C)},$$

$$u_{Sn} = 4u1_{(Sn)} + 6u2_{(Sn)} + u3_{(Sn)},$$

The strain energies u_C and u_{Sn} are the function of two corresponding variables d and l, and their values are obtained from the minimization of the internal strain energy. The internal strain energy due to isolated carbon and Sn atoms, respectively, is [11]:

$$u_C = 172 \text{ kJ mol}^{-1}, \quad u_{Sn} = 13.61 \text{ kJ mol}^{-1}.$$

The bond-stretching and bond-bending elastic constants used in the calculations were taken from Refs [1,12].

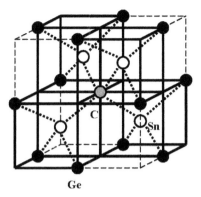

FIGURE 6.16 Quadruple of 1C3Ge1Sn tetrahedral cells around the central carbon atom.

The isoelectronic impurity tetrahedral 4Sn1C cluster shown in Figure 6.16 forms the quadruple of 1C3Ge1Sn tetrahedral cells around the central carbon atom.

The distortions of the crystal lattice and the internal strain energy due to the cluster are obtained in a similar way to the cases of the isolated impurities. This cluster forms four 1C3Ge1Sn tetrahedral cells around the central carbon atom. The deformation energies of 1C3Ge1Sn tetrahedral cells are the first item in the sum representing the internal strain energy caused by the cluster. The deformation energy of 1C3Ge1Sn tetrahedral cell $u_{1(4Sn1C)}$ is estimated as follows. The carbon atom is at the origin of coordinates (Figure 6.17).

The central atom in the 4Sn1C cell is at coordinates $x = y = z = l$. The size of the edge of the cube with the tetrahedral cell is equal to d, and there are three of the same SnGe(4Sn1C) bonds and one CSn(4Sn1C) bond. There are three of the same CSnGe(4Sn1C) and three of the same GeSn-Ge(4Sn1C) angles between the bonds in the tetrahedral cell. The lengths of

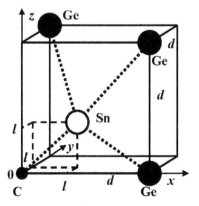

FIGURE 6.17 1C3Ge1Sn tetrahedral cell.

the bonds and the cosines of the angles between the bonds are given, respectively, by:

$$r_{CSn(4Sn1C)} = \sqrt{3}\, l,$$

$$r_{SnGe(4Sn1C)} = \sqrt{3l^2 - 4dl + 2d^2},$$

$$\cos \varphi_{CSnGe(4Sn1C)} = \frac{3l - 2d}{\sqrt{3}\sqrt{l^2 + 2(d-l)^2}},$$

$$\cos \varphi_{GeSnGe(4Sn1C)} = \frac{-2l(d-l) + (d-l)^2}{\sqrt{2(d-l)^2 + l^2}\sqrt{2(d-l)^2 + l^2}},$$

The deformation energies of the bonds and the angles between the bonds are, respectively:

$$u_{CSn(4Sn1C)} = \frac{3(\alpha_C + \alpha_{Sn})}{4(R_C + R_{Sn})^2}\left[\left(\frac{R_C + R_{Sn}}{2}\right)^2 - 3l^2\right]^2,$$

$$u_{SnGe(4Sn1C)} = \frac{3(\alpha_{Sn} + \alpha_{Ge})}{4(R_{Sn} + R_{Ge})^2}\left[\left(\frac{R_{Sn} + R_{Ge}}{2}\right)^2 - 3l^2 + 4dl - 2d^2\right]^2,$$

$$u_{CSnGe(4Sn1C)} = \frac{3(\alpha_C + 2\alpha_{Sn} + \alpha_{Ge})}{4(R_C + R_{Sn})(R_{Sn} + R_{Ge})}$$
$$\left[-\frac{(R_C + R_{Sn})(R_{Sn} + R_{Ge})}{12}\right.$$
$$\left. - l\sqrt{3l^2 - 4dl + 2d^2}\,\frac{3l - 2d}{\sqrt{l^2 + 2(d-l)^2}}\right]^2,$$

$$u_{GeSnGe(4Sn1C)} = \frac{3(\beta_{Sn} + \beta_{Ge})}{4(R_{Sn} + R_{Ge})^2}$$
$$\left[-\frac{1}{3}\left(\frac{R_{Sn} + R_{Ge}}{2}\right)^2\right.$$
$$\left. - l\sqrt{3l^2 - 4dl + 2d^2}\,\frac{3l - 2d}{\sqrt{l^2 + 2(d-l)^2}}\right]^2,$$

where the bond-stretching elastic constant of the CSn(4Sn1C) bond is equal to $\frac{\alpha_C + \alpha_{Sn}}{2}$, the lengthd of the undistorted CSn(4Sn1C) and SnGe(4Sn1C) bonds are equal, respectively, to $\frac{R_C + R_{Sn}}{2}$ and $\frac{R_{Sn} + R_{Ge}}{2}$, and the bond-bending elastic constants of CSnGe(4Sn1C) and GeSnGe(4Sn1C) angles are $\frac{\beta_C + 2\beta_{Sn} + \beta_{Ge}}{4}$ and $\frac{\beta_{Sn} + \beta_{Ge}}{2}$, respectively. The strain energy of the 1C3Ge1Sn tetrahedral cell is:

$$u_{1(4Sn1C)} = u_{CSn(4Sn1C)} + 3u_{SnGe(4Sn1C)} + 3u_{CSnGe(4Sn1C)} + 3u_{GeSnGe(4Sn1C)}.$$

The angles belonging to the different tetrahedral cells in the quadruples are undistorted, as seen in Figure 6.14. Accordingly, the deformation energy of the SnCSn(4Sn1C) angle belonging to the different 1C3Ge1Sn tetrahedral cells $u_{2(4Sn1C)}$ is:

$$u_{2(4Sn1C)} = \frac{\beta_C + \beta_{Sn}}{12\left(\frac{R_{Sn} + R_{Ge}}{2}\right)^2}\left[\left(\frac{R_{Sn} + R_{Ge}}{2}\right)^2 - 3l^2\right]^2.$$

The strain energy of the crystal lattice outside the quadruple of the 1C3Ge1Sn tetrahedral cells is derived similarly to the strain energy of the crystal lattice outside the other quadruples:

$$u_{3(4Sn1C)} = \frac{16\sqrt{2}}{5}\pi(C_{11} - C_{12} + 3C_{44})d\left(d - \frac{2}{\sqrt{3}}R_{Ge}\right)^2,$$

where C_{11} is the stiffness coefficient of Ge.

The internal strain energy due to the 4Sn1C tetrahedral cluster is the function of two corresponding variables, l and d, given by:

$$u_{4Sn1C} = 4u_{1(4Sn1C)} + 6u_{2(4Sn1C)} + u_{3(4Sn1C)}$$

and it is obtained by minimization.

The internal strain energy due to the 4Sn1C clusters is [13]:

$$u_{4Sn1C} = 17.3 \text{ kJ mol}^{-1}.$$

The bond-stretching and bond-bending elastic constants used in the calculations are available in Refs [1,12].

The estimated internal strain energies due to the isolated carbon and Sn atoms and 4Sn1C clusters in Ge doped with carbon and Sn demonstrate the significant compensation of the internal strains after the formation of the clusters. Therefore, the isolated carbon and Sn atoms in such alloys have the tendency to self-assembling of 4Sn1C clusters.

6.8 INTERNAL STRAIN ENERGY OF BINARY COMPOUNDS DUE TO ISOELECTRONIC IMPURITY

Isoelectronic impurities in binary semiconductor compounds are, normally, over the sublattice sites corresponding to the group of the periodic table of the host atoms. Moreover, isoelectronic impurity atoms form chemical bonds with the nearest neighbors situated over another sublattice sites. Hence, doping with isoelectronic impurity transforms a binary compound in a ternary substitutional alloy of two binary compounds. A mismatched isoelectronic impurity dilates or shrinks the crystal lattice of a semiconductor matrix around itself. The distortions of the crystal lattice and the internal strain energy due to the isoelectronic impurity in the binary semiconductor compound are represented here.

Let the binary compound AB with the zinc blende structure be doped with the isoelectronic anions C forming the ternary alloy AC_xB_{1-x} of AB and AC binary compounds. It is supposed that the concentration of the isoelectronic impurity satisfies the condition $x \ll 1$. The consideration of the binary compound doped with isoelectronic cations is done in a similar way. The isolated impurities form around themselves the quadruples of $1C3A1B$ ($R_{CB} \geq R_{AB}$) or $3A1C1B$ ($R_{AB} > R_{CB}$) tetrahedral cells shown in Figures 6.18 and 6.19, respectively, where R_{CB} is the distance between the nearest neighbors in the undistorted compound CB. The notations of the tetrahedral cells correspond to the notations introduced in Section 6.2, in which on the first place there is the atom of the compound with the larger lattice parameter.

The distortions of the crystal lattice and the internal strain energy due to the isoelectronic impurity anion C are described in a way similar to that used when considering the carbon and Sn impurities in Ge represented in Section 6.8. The Helmholtz free energy of AC_xB_{1-x} alloy is:

$$f = f_{AB}^0 + f_{AC}^0 + u^{\text{Int.Str.}} - Ts^{\text{Conf.}},$$

where f_{AB}^0, $u^{\text{Int.Str.}}$, and $s^{\text{Conf.}}$ are the free energy of the binary compound AB, the internal strain energy, and the configurational entropy, respectively. The distortions of the crystal lattice and the internal strain energy

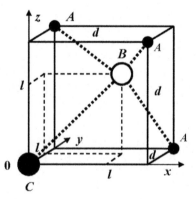

FIGURE 6.18 $1C3A1B$ tetrahedral cell.

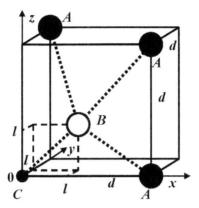

FIGURE 6.19 3A1C1B tetrahedral cell.

are obtained providing that the internal strain energy is represented as a sum of three energies. The deformation energy of the tetrahedral cells is the first item in the sum. Due to the symmetry of the zinc blende structure, the tetrahedral cells have the shape of the geometrically undistorted tetrahedrons of the same size. The deformation energies of a bond and an angle between bonds by using the valence force field model, used here before, are given in expressions (6.7.1) and (6.7.2). Each tetrahedral cell contains four bonds and six angles between them.

The deformation energy of a 1C3A1B ($R_{CB} > R_{AB}$) tetrahedral cell $u_{1(1C3A1B)}$ can be estimated as follows. Atom C is at the origin of coordinates (Figures 6.18 and 6.19). Atom B is at coordinates $x = y = z = l$. The size of the edge of the cube with the tetrahedral cell is equal to d. There are three of the same AB bonds and one AC bond. There are three of the same CBA and three of the same ABA angles in the 1C3A1B tetrahedral cell. The lengths of the bonds and the angles are given, respectively, by;

$$r_{CB} = \sqrt{3}l,$$

$$r_{AB} = \sqrt{3l^2 - 4dl + 2d^2},$$

$$\cos \varphi_{CBA} = \frac{3l - 2d}{\sqrt{3}\sqrt{l^2 + 2(d - l)^2}},$$

$$\cos \varphi_{ABA} = \frac{-2l(d - l) + (d - l)^2}{\sqrt{2(d - l)^2 + l^2}\sqrt{2(d - l)^2 + l^2}},$$

The angles between bonds are obtained from the formula for the cosine of the angle α between the vectors $\mathbf{a} = (x_1, y_1, z_1)$ and $\mathbf{b} = (x_2, y_2, z_2)$ known from analytic geometry (Eqn (6.7.3)). The deformation energies of the bonds and angles are written, respectively, as:

$$u_{CB} = \frac{3\alpha_{CB}}{8R_{CB}^2}\left(R_{CB}^2 - 3l^2\right)^2,$$

$$u_{AB} = \frac{3\alpha_{AB}}{8R_{AB}^2}\left(R_{AB}^2 - 3l^2 + 4dl - 2d^2\right)^2,$$

$$u_{CBA} = \frac{3(\beta_{CB} + \beta_{AB})}{16R_{CB}R_{AB}}\left[-\frac{1}{3}R_{CB}R_{AB} - \sqrt{3l^2 - 4dl + 2d^2}\;\frac{l^2 - 2l(d-l)}{\sqrt{l^2 + 2(d-l)^2}}\right]^2,$$

$$u_{ABA} = \frac{3\beta_{AB}}{8R_{AB}^2}\left[-\frac{1}{3}R_{AB}^2 - \left(3l^2 - 4dl + 2d^2\right)\frac{3l^2 - 4dl + d^2}{l^2 + 2(d-l)^2}\right]^2.$$

The deformation energy of 1C3A1B tetrahedral cell $u_{1(1C3A1B)}$ is:

$$u_{1(1C3A1B)} = u_{CB} + 3u_{AB} + 3u_{CBA} + 3u_{ABA}.$$

The deformation energy of a 3A1C1B ($R_{CB} < R_{AB}$) tetrahedral cell shown in Figure 6.14b is estimated in a similar way. The atom C is at the origin and the atom B is at coordinates $x = y = z = l$. The size of the edge of the cube with the tetrahedral cell is equal to d. There are three of the same AB bonds and one CB bond, and there are three of the same CBA and three of the same ABA angles. The lengths of the bonds and the angles between the bonds are given, respectively, by:

$$r_{AB} = \sqrt{3l^2 - 4dl + 2d^2},$$

$$r_{CB} = \sqrt{3}l,$$

$$\cos\varphi_{CBA} = \frac{3l - 2d}{\sqrt{3}\sqrt{l^2 + 2(d-l)^2}},$$

$$\cos\varphi_{ABA} = \frac{3l^2 - 4dl + d^2}{l^2 + 2(d-l)^2}.$$

The deformation energies of the bonds and the angles between bonds are, correspondingly:

$$u_{AB} = \frac{3\alpha_{AB}}{8R_{AB}^2}\left(R_{AB}^2 - 3l^2 + 4dl - 2d^2\right)^2,$$

$$u_{CB} = \frac{3\alpha_{CB}}{8R_{CB}^2}\left(R_{CB}^2 - 3l^2\right)^2,$$

$$u_{CBA} = \frac{3(\beta_{CB} + \beta_{AB})}{16R_{CB}R_{AB}} \left[-\frac{1}{3}R_{CB}R_{AB} - \sqrt{3l^2 - 4dl + 2d^2} \; \frac{l^2 - 2l(d-l)}{\sqrt{l^2 + 2(d-l)^2}} \right]^2,$$

$$u_{ABA} = \frac{3\beta_{AB}}{8R_{AB}^2} \left[-\frac{1}{3}R_{AB}^2 - (3l^2 - 4dl + 2d^2) \frac{3l^2 - 4dl + d^2}{l^2 + 2(d-l)^2} \right]^2.$$

The deformation energy of $3A1C1B$ tetrahedral cell $u_{1(3A1C1B)}$ is:

$$u_{1(3A1C1B)} = u_{CB} + 3u_{AB} + 3u_{CBA} + 3u_{ABA}.$$

The deformation energies of the angles between the bonds belonging to the different tetrahedral cells in the quadruples are the second items of the sums representing the energies due to the impurity C in both types of the tetrahedral cells. The angles between the bonds are undistorted, as it is seen from Figure 6.14. Accordingly, the deformation energies of the angles BCB belonging to the different tetrahedral cells in the quadruples $u_{2(1C3A1B)}$ and $u_{2(3A1C1B)}$ are, respectively:

$$u_{2(1C3A1B)} = \frac{\beta_{CB}}{24R_{CB}^2}\left(R_{CB}^2 - 3l^2\right)^2, \quad \text{where } l = l_{(1C3A1B)},$$

$$u_{2(3A1C1B)} = \frac{\beta_{CB}}{24R_{CB}^2}\left(R_{CB}^2 - 3l^2\right)^2, \quad \text{where } l = l_{(3A1C1B)}.$$

The strain energies of the crystal lattice outside the quadruples are the third items of the sums. These energies are considered as the strain energies of the crystal lattices outside the spheres centered at the impurity atoms due to the symmetry of the quadruples in the zinc blende structure. The radia of the spheres are the diagonals of the faces of the cubes with the tetrahedron cells equal to $\sqrt{2}d$ (Figure 6.14) with the origin of coordinates at the impurity atoms. The radial displacement of the atoms of the crystal lattice outside the quadruple at the distance r is, as well as in Section 6.7, supposed expressed as $u \propto r^{-2}$ similarly to the displacement in an elastic medium [10]. The projections of the vector $\mathbf{r}$ on the coordinate axes are denoted by r_x, r_y and r_z. If the displacement of $\sqrt{2}d$ is:

$$u\left(\sqrt{2}d\right) = \sqrt{2}\left(d - \frac{2}{\sqrt{3}}R_{AB}\right),$$

where R_{AB} is the distance between the nearest neighbors in the undistorted compound AB, then:

$$u(r) = u\left(\sqrt{2}d\right)\frac{2d^2}{r^2}.$$

The strain energy density (Eqn (6.7.4)) is:

$$\varepsilon = \frac{1}{2}C_{11}(e_{xx} + e_{yy} + e_{zz})^2 - (C_{11} - C_{12})(e_{xx}e_{yy} + e_{yy}e_{zz} + e_{zz}e_{xx})$$
$$+ \frac{1}{2}C_{44}\left(e_{xy}^2 + e_{yz}^2 + e_{zx}^2\right),$$

where C_{11} is the stiffness coefficient of the compound AB, and e_{ij} ($i, j = x, y, z$) are the components of the strain tensor. The components of the strain tensor of the radially strained elastic medium in the spherical coordinate system obtained in Section 6.7 are described by expressions (6.7.5–6.7.10).

The strain energies of the crystal lattices outside the quadruples (Eqn (6.7.11)) are:

$$u_3 = \int_{\sqrt{2}d}^{\infty} \varepsilon(r)d^3r = \frac{16\sqrt{2}}{5}\pi(C_{11} - C_{12} + 3C_{44})d\left(d - \frac{2}{\sqrt{3}}R_{AB}\right)^2,$$

where $u_3 = u_{3(1A3C1B)}$ or $u_{3(3A1C1B)}$, C_{11} is the stiffness coefficient of the compound AB, and $d = d_{(1C3A1B)}$ or $d_{(3A1C1B)}$.

The total internal strain energies due to the mismatched isoelectronic impurity anions C in the binary compound AB are given by:

$$u_{1A3C1B} = 4u_{1(1A3C1B)} + 6u_{2(1A3C1B)} + u_{3(1A3C1B)}, \quad (R_{CB} > R_{AB}),$$

$$u_{3A1C1B} = 4u_{1(3A1C1B)} + 6u_{2(3A1C1B)} + u_{3(3A1C1B)}, \quad (R_{CB} > R_{AB}).$$

These energies are the function of two variables, $d_{(1C3A1B)}$ or $d_{(3A1C1B)}$ and $l_{(1C3A1B)}$ or $l_{(3A1C1B)}$, and they are obtained by using their minimizing.

References

[1] P.N. Keating, Effect of invariance requirements on the elastic strain energy of crystals with applications to the diamond structure, Phys. Rev. 145 (2) (1966) 637–645.

[2] R.M. Martin, Elastic properties of ZnS structure semiconductors, Phys. Rev. B 1 (10) (1970) 4005–4010.

[3] R.M. Martin, Relation between elastic tensors of wurtzite and zinc-blende structure materials, Phys. Rev. B 6 (12) (1972) 4546–4553.

[4] T. Fukui, Calculation of bond length in $Ga_{1-x}In_xAs$ ternary semiconductors, Jpn. J. Appl. Phys. Part 2 23 (4) (1984) L208–L211.

[5] J.C. Mikkelsen Jr., J.B. Boyce, Atomic-scale structure of random solid solutions: extended X-ray-absorption fine-structure study of $Ga_{1-x}In_xAs$, Phys. Rev. Lett. 49 (19) (1982) 1412–1415.

[6] J.C. Mikkelsen Jr., The atomic-scale origin of the strain enthalpy of mixing in zincblende alloys, J. Electrochem. Soc. 132 (2) (1985) 500–505.

[7] A. Balzarotti, A. Kisiel, N. Motta, M. Zimnal-Starnawska, M.T. Czyzyk, M. P Podgorny, Local structure of ternary semiconducting random solid solutions: extended X-ray-absorption fine structure of $Cd_{1-x}Mn_xTe$, Phys. Rev. B 30 (4) (1984) 2295–2298.

[8] V. Kachkanov, K.P. O'Donnell, R.W. Martin, J.F.W. Mosselmans, S. Pereira, Local structure of luminescent InGaN alloys, Appl. Phys. Lett. 89 (10) (2006) 101908.

[9] J.G. Thompson, R.L. Withers, S.R. Pelethhorpe, A. Melnitchenko, Crystaballite-related oxide structures, J. Solid State Chem. 141 (1) (1998) 29–49.

[10] L.D. Landau, E.M. Lifshitz, Theory of Elasticity, in: Course of Theoretical Physics, second ed., vol. 7, Pergamon Press, Oxford, 1975 (Chapter 1).

[11] V.A. Elyukhin, S.F. Diaz Albarran, Self-assembling of C and Sn in Ge, Physica E 28 (4) (2005) 552–555.

[12] J.L. Martins, A. Zunger, Bond lengths around isovalent impurities and in semiconductor solid solutions, Phys. Rev. B 30 (10) (1984) 6217–6220.

[13] V.A. Elyukhin, L.P. Sorokina, V.A. Mishurnyi, F. de Anda, Self-assembling of 4C10Sn clusters in Ge co-doped with C and Sn, Physica E 40 (4) (2008) 883–885.

Strain Energy of Thin Lattice-Mismatched Layers in Crystals with Cubic Structure

To find the interface orientation ensuring the minimal strain energy of the elastically strained lattice-mismatched layer, the strain energy of the arbitrarily oriented layer should be described. The strain energy of the thin lattice-mismatched layer and the bulk crystal having the cubic structure is regarded as follows. The consideration is strongly influenced by the use of the model of Hornstra and Bartels [1]. It is assumed that the thickness of the layer is much smaller than the dimensions of the bulk crystal. In such a case, the strain energy of the layer and the bulk crystal is only the strain energy of the layer. Moreover, it is supposed that the curvature of the layer is absent due to its small thickness. The coordinates and the displacements of atoms of the mismatched layer are taken along the crystallographic axes and the origin of coordinates is taken in the interfacial plane (Figure A1.1).

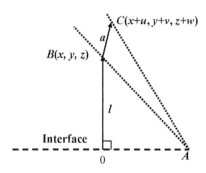

FIGURE A1.1 The cross-section perpendicular to the interface, the atomic plane AB of the bulk crystal and the atomic plane AC of the layer.

These displacements are the result of the movement of the atoms in the layer from the strained state (shown by the AB line in Figure A1.1), when its structure was strained in all directions (x,y,z) by:

$$\varepsilon_i = -\frac{a_{L,i} - a_{BC,i}}{a_{L,i}}, (i = x, y, z),$$

where $a_{L,i}$ and $a_{BC,I}$ are the lattice parameters of the layer and the bulk crystal in the i-th direction, respectively, to the relaxed state (shown by the AC line in Figure A1.1). The AB line is the atomic plane of the bulk crystal. These displacements of all atoms in the layer are equal. There are no stresses at the interface or the components of the force at the interface vanish, which is written as:

$$f_x = l_1\sigma_1 + l_2\sigma_6 + l_3\sigma_5 = 0, \tag{A1.1}$$

$$f_y = l_1\sigma_6 + l_2\sigma_2 + l_3\sigma_4 = 0, \tag{A1.2}$$

$$f_z = l_1\sigma_5 + l_2\sigma_4 + l_3\sigma_3 = 0, \tag{A1.3}$$

where l_1, l_2, and l_3 are the direction cosines of the interface normal l (Figure A1.1) and σ_i $(i = 1,\ldots,6)$ are the components of the stresses.

For cubic crystals, the components of the stresses σ_i $(i = 1,\ldots,6)$ and the strains ε_j $(j = 1,\ldots,6)$ in the contracted notation are related by:

$$
\begin{pmatrix} \sigma_1 \\ \sigma_2 \\ \sigma_3 \\ \sigma_4 \\ \sigma_5 \\ \sigma_6 \end{pmatrix}
=
\begin{pmatrix}
C_{11} & C_{12} & C_{12} & 0 & 0 & 0 \\
C_{12} & C_{11} & C_{12} & 0 & 0 & 0 \\
C_{12} & C_{12} & C_{11} & 0 & 0 & 0 \\
0 & 0 & 0 & C_{44} & 0 & 0 \\
0 & 0 & 0 & 0 & C_{44} & 0 \\
0 & 0 & 0 & 0 & 0 & C_{44}
\end{pmatrix}
\begin{pmatrix} \varepsilon_1 \\ \varepsilon_2 \\ \varepsilon_3 \\ \varepsilon_4 \\ \varepsilon_5 \\ \varepsilon_6 \end{pmatrix}
\quad \text{or}
$$

$$\sigma_1 = \sigma_2 = (C_{11} + C_{12})\varepsilon_1 + C_{12}\varepsilon_3,$$

$$\sigma_3 = 2C_{12}\varepsilon_1 + C_{11}\varepsilon_3,$$

$$\sigma_4 = C_{44}\varepsilon_4, \quad \sigma_5 = C_{44}\varepsilon_5, \quad \sigma_6 = C_{44}\varepsilon_6,$$

where (C_{ij}) is the matrix of the stiffness coefficients in the contracted notation for the cubic crystals in the crystallographic system of the coordinates.

The layer lattice displacements u, v, and w (C in Figure A1.1) are related to the coordinates x, y, and z (B in Figure A1.1) by the strain tensor A_{ij} as follows:

$$u = A_{11}x + A_{12}y + A_{13}z,$$

$$v = A_{21}x + A_{22}y + A_{23}z,$$

$$w = A_{31}x + A_{32}y + A_{33}z,$$

where $A_{ij} = a_i l_j + \varepsilon_\| \delta_{ij}$, the vector a is the vector equal to the displacement of any lattice point in the layer, divided by the distance of that lattice point from the interface, $\varepsilon_\| = -\frac{a_L - a_{BC}}{a_L}$ is the strain of the layer parallel to the interface, and a_L and a_{BC} are the lattice parameters of the layer and the bulk crystal, respectively. The relations between the components of the strain tensor and the components of the strain are:

$$\varepsilon_1 = A_{11}, \quad \varepsilon_2 = A_{22}, \quad \varepsilon_3 = A_{33}, \quad \varepsilon_4 = A_{23} + A_{32},$$

$$\varepsilon_5 = A_{13} + A_{31}, \quad \varepsilon_6 = A_{12} + A_{21}.$$

Finally, Eqns (A1.1–A1.3) can be rewritten as:

$$a_1 \left[(C_{11} - C_{44})l_1^2 + C_{44} \right] + a_2 l_1 l_2 (C_{12} + C_{44})$$
$$+ a_3 l_1 l_3 (C_{12} + C_{44}) + \varepsilon_\| l_1 (C_{11} + 2C_{12}) = 0, \tag{A1.4}$$

$$a_1 l_1 l_2 (C_{12} + C_{44}) + a_2 \left[(C_{11} - C_{44})l_2^2 + C_{44} \right]$$
$$+ a_3 l_2 l_3 (C_{12} + C_{44}) + \varepsilon_\| l_2 (C_{11} + 2C_{12}) = 0, \tag{A1.5}$$

$$a_1 l_1 l_3 (C_{44} + C_{12}) + a_2 (C_{44} l_2 l_3 + C_{12} l_2 l_3)$$
$$+ a_3 \left[(C_{11} - C_{44})l_3^2 + C_{44} \right] + \varepsilon_\| l_3 (C_{11} + 2C_{12}) = 0, \tag{A1.6}$$

and allow us to express two components a_2 and a_3 of the vector a as functions of its third component a_1, which are given by:

$$a_2 = a_1 \frac{l_2 \left[C_{44} - l_1^2 (2C_{44} - C_{11} + C_{12}) \right]}{l_1 \left[C_{44} - l_2^2 (2C_{44} - C_{11} + C_{12}) \right]}, \tag{A1.7}$$

$$a_3 = a_1 \frac{l_3 \left[C_{44} - l_1^2 (2C_{44} - C_{11} + C_{12}) \right]}{l_1 \left[C_{44} - l_3^2 (2C_{44} - C_{11} + C_{12}) \right]}. \tag{A1.8}$$

The third component, a_1, is calculated from one of Eqns (A1.4–A1.6) by using Eqns (A1.7) and A1.8).

The strain energy of the insignificant lattice-mismatched layer is given as:

$$u^{SE} = \frac{1}{2} v_{BC} \sum_{i=1}^{6} \sigma_i \varepsilon_i,$$

where v_{BC} is the molar volume of the bulk crystal. The estimates fulfilled for the semiconductors demonstrate that the strain energy of the layer attains the minimum value, if only one of the direction cosines of the interface normal l is not equal to zero, or is in the (001), (010), and (100) lattice planes of the bulk crystal. The strain energy of the layer in the (001), (010), and (100) lattice planes is written as:

$$u^{SE} = \frac{1}{2}(\sigma_1 \varepsilon_1 + 2\sigma_2 \varepsilon_2) v_{BC},$$

where $\varepsilon_2 = \varepsilon_\parallel = \frac{a_L - a_{BC}}{a_{BC}}$. The axis perpendicular to the layer–bulk crystal interface is characterized by vanishing forces in the direction $\sigma_1 = 0$. Therefore, the stresses parallel to the layer–bulk crystal interface are given by:

$$\sigma_2 = \frac{(C_{11} - C_{12})(C_{11} + 2C_{12})}{C_{11}} \frac{a_L - a_{BC}}{a_{BC}},$$

where C_{11} is the stiffness coefficient of the material of the layer. Accordingly, the strain energy is:

$$u^{SE} = \frac{(C_{11} - C_{12})(C_{11} + 2C_{12})}{C_{11}} v_{BC} \left(\frac{a_L - a_{BC}}{a_{BC}}\right)^2.$$

Reference

[1] J. Hornstra, W.J. Bartels, Determination of the lattice constant of epitaxial layers of III-V compounds, J. Cryst. Growth 44 (4) (1978) 513–517.

Strain Energy of Thin Lattice-Mismatched Layers in Crystals with a Wurtzite Structure

To find the interface orientation ensuring the minimal strain energy of the elastically strained lattice-mismatched layer in a bulk crystal, the strain energy of the arbitrarily oriented layer should be described. The strain energy of the thin lattice-mismatched layer with the wurtzite structure in a bulk crystal with the same structure is represented as follows. The consideration is developed by using the model from Ref. [1]. It is assumed that the thickness of the layer is much smaller then the dimensions of the bulk crystal. In such a case, the strain energy is the strain energy of the layer only. Moreover, it is supposed that the curvature of the layer is absent due to its small thickness. The coordinates and the displacements of atoms of the mismatched layer are taken along the crystallographic axes, and the origin of coordinates is in the interfacial plane (Figure A2.1).

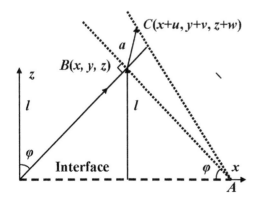

FIGURE A2.1 Cross-section perpendicular to the interface, the atomic plane AB of the bulk crystal, and atomic plane AC of the layer.

These displacements are the result of the movement of the atoms in the layer from the state when it is strained in all directions (x,y,z) by:

$$\varepsilon_i = -\frac{a_{L,i} - a_{BC,i}}{a_{L,i}}, \quad (i = x, y, z),$$

where $a_{L,i}$ and $a_{BC,I}$ are the lattice parameters of the layer and the bulk crystal in the i-th direction, respectively, to the relaxed state. These displacements of all atoms in the layer are equal. There are no stresses on the interfaces that is written as:

$$f_x = l_1\sigma_1 + l_2\sigma_6 + l_3\sigma_5 = 0,$$
$$f_y = l_1\sigma_6 + l_2\sigma_2 + l_3\sigma_4 = 0,$$
$$f_z = l_1\sigma_5 + l_2\sigma_4 + l_3\sigma_3 = 0,$$

where l_1, l_2, and l_3 are the direction cosines of the interface normal l (Figure A2.1). For hexagonal crystals according to Hermann's theorem: "If an r-rank tensor has an N-fold symmetry axis and $r < N$, then this tensor also has a symmetry axis of an infinite order" [2]. The strain energy is isotropic in the <0001> zone, since the stiffness coefficients form a fourth-rank tensor that is too low to resolve the six-fold symmetry axis. Therefore, the lattice-mismatched layer is chosen to be parallel to the x axis to simplify the calculations. As a result, the strain energy of the layer depends only on the angle φ between the z axis and the interface normal l, since the AB line is the atomic plane of the bulk crystal. This angle is in interval from 0 to $\pi/2$ and $l_1 = 0$, $l_2 = \sin\varphi$, $l_3 = \cos\varphi$.

The components of the stresses σ_i $(i = 1,...,6)$ and strains ε_j $(j = 1,...,6)$ in the contracted notation are related by the expressions:

$$\begin{pmatrix} \sigma_1 \\ \sigma_2 \\ \sigma_3 \\ \sigma_4 \\ \sigma_5 \\ \sigma_6 \end{pmatrix} = \begin{pmatrix} C_{11} & C_{12} & C_{13} & 0 & 0 & 0 \\ C_{12} & C_{11} & C_{13} & 0 & 0 & 0 \\ C_{13} & C_{13} & C_{33} & 0 & 0 & 0 \\ 0 & 0 & 0 & C_{44} & 0 & 0 \\ 0 & 0 & 0 & 0 & C_{44} & 0 \\ 0 & 0 & 0 & 0 & 0 & C_{66} = \frac{1}{2}(C_{11} - C_{12}) \end{pmatrix} \times \begin{pmatrix} \varepsilon_1 \\ \varepsilon_2 \\ \varepsilon_3 \\ \varepsilon_4 \\ \varepsilon_5 \\ \varepsilon_6 \end{pmatrix} \quad \text{or}$$

$$\sigma_1 = C_{11}\varepsilon_1 + C_{12}\varepsilon_2 + C_{13}\varepsilon_3,$$
$$\sigma_2 = C_{12}\varepsilon_1 + C_{11}\varepsilon_2 + C_{13}\varepsilon_3,$$
$$\sigma_3 = C_{13}(\varepsilon_1 + \varepsilon_2) + C_{33}\varepsilon_3,$$
$$\sigma_4 = C_{44}\varepsilon_4, \quad \sigma_5 = C_{44}\varepsilon_5, \quad \sigma_6 = C_{66}\varepsilon_6.$$

where (C_{ij}) is the matrix of the stiffness coefficients for the hexagonal crystals in the crystallographic system of the coordinates.

The layer lattice displacements u, v, and w (Figure A2.1) are related to the coordinates x, y, and z by the strain tensor A_{ij} as follows:

$$u = A_{11}x + A_{12}y + A_{13}z,$$

$$v = A_{21}x + A_{22}y + A_{23}z,$$

$$w = A_{31}x + A_{32}y + A_{33}z.$$

The relations between the components of the strain tensor A_{ij} and strains ε_i are:

$$\varepsilon_1 = A_{11}, \quad \varepsilon_2 = A_{22}, \quad \varepsilon_3 = A_{33},$$

$$\varepsilon_4 = A_{23} + A_{32}, \quad \varepsilon_5 = A_{13} + A_{31}, \quad \varepsilon_6 = A_{12} + A_{21}.$$

The components of the strain tensor are:

$$A_{ij} = a_i l_j + \varepsilon_{ij},$$

where a_i refers to the components of the vector a equal to the distance between the position of an atom in the layer and the position of the same atom in the lattice corresponding to the bulk crystal, and ε_{ij} is the change in the lattice parameters across the interface. The relations between the components of the strain tensor A_{ij} and strains ε_i, which are the contracted form of the tensor, are given by:

$$\varepsilon_1 = A_{11}, \quad \varepsilon_2 = A_{22},$$

$$\varepsilon_3 = A_{33}, \quad \varepsilon_4 = A_{23} + A_{32},$$

$$\varepsilon_5 = A_{13} + A_{31}, \quad \varepsilon_6 = A_{12} + A_{21}.$$

The stresses σ_i and strains ε_j of the insignificantly lattice mismatched layer are related by the stiffness coefficients C_{ij} as $\sigma_i = C_{ij}\varepsilon_j$ in the contracted tensor notation. The stresses are, respectively:

$$\sigma_1 = C_{11}\varepsilon_a + C_{12}(a_2 l_2 + \varepsilon_a) + C_{13}(a_3 l_3 + \varepsilon_c), \tag{A2.1}$$

$$\sigma_2 = C_{12}\varepsilon_a + C_{11}(a_2 l_2 + \varepsilon_a) + C_{13}(a_3 l_3 + \varepsilon_c), \tag{A2.2}$$

$$\sigma_3 = C_{13}(a_2 l_2 + 2\varepsilon_a) + C_{33}(a_3 l_3 + \varepsilon_c), \tag{A2.3}$$

$$\sigma_4 = C_{44}(a_2 l_3 + a_3 l_2), \tag{A2.4}$$

$$\sigma_5 = C_{44}a_1 l_3, \tag{A2.5}$$

$$\sigma_6 = C_{66}a_1 l_2, \tag{A2.6}$$

where C_{11} is the stiffness coefficient of the material of the layer, $\varepsilon_a \equiv \varepsilon_{11} = \varepsilon_{22} = -\left(\frac{a_L - a_{BC}}{a_{BC}}\right)$, $\varepsilon_c \equiv \varepsilon_{33} = -\left(\frac{c_L - c_{BC}}{c_{BC}}\right)$, a_L, c_L, and a_{BC}, c_{BC} are the lattice parameters of the layer and the bulk crystal, respectively. There are no stresses on the interface, and therefore:

$$l_2\sigma_6 + l_3\sigma_5 = 0, \tag{A2.7}$$

$$l_2\sigma_2 + l_3\sigma_4 = 0, \tag{A2.8}$$

$$l_2\sigma_4 + l_3\sigma_3 = 0. \tag{A2.9}$$

The strain energy of the insignificantly lattice-mismatched layer is:

$$u = \frac{1}{2} v_{BC} \sum_{i=1}^{6} \sigma_i \varepsilon_i,$$

where v_{BC} is the molar volume of the bulk crystal. Formulas (A2.1–A2.6), Eqns (A2.7–A2.9), and the condition:

$$l_1 = 0, \quad l_2 = \sin \varphi, \quad l_3 = \cos \varphi$$

allow us to describe the components a_i as functions of the one-direction cosine of the interface normal. Finally, the strain energy is obtained by its minimizing. The orientation of the layer ensuring the minimum of the strain energy depends on the constituents and the composition of a semiconductor alloy.

References

[1] D.J. Bottomley, P. Fons, D.J. Tweet, Determination of the lattice constants of epitaxial layers, J. Cryst. Growth 154 (3–4) (1995) 401–409.
[2] Yu I. Sirotin, M.P. Shaskolskaya, Fundamentals Crystal Physics, Mir Publishers, Moscow, 1982 (Chapter V).

3

Method of Lagrange Undetermined Multipliers

Let F be a function of n variables x_i $(i = 1,...,n)$. The function F attains its extremum (maximum or minimum value) if the following condition:

$$dF = \sum_{i=1}^{n} \frac{\partial F}{\partial x_i} dx_i = 0 \qquad (A3.1)$$

is fulfilled. If all variables are independent, then this condition can be rewritten as:

$$\frac{\partial F}{\partial x_i} = 0, \quad (i = 1, ..., n).$$

If the variables are reciprocally dependent, or in other words connected by $m < n$ constraints given by:

$$\varphi_j(x_i) = 0, \quad (j = 1, ..., m), \qquad (A3.2)$$

then m differentials dx_i in Eqn (A3.1) should be consistent with Eqn (A3.2), so that the conditions:

$$\sum_{i=1}^{n} \frac{\partial \varphi_j}{\partial x_i} dx_i = 0 \qquad (A3.3)$$

are fulfilled. The conditions (A3.3) may be rewritten as:

$$\sum_{i=1}^{n} \lambda_j \frac{\partial \varphi_j}{\partial x_i} dx_i = 0, \qquad (A3.4)$$

where λ_j are functions of the variables x_i called the Lagrange undetermined multipliers. The conditions (A3.1) and (A3.4) can be rewritten as a condition given by:

$$\sum_{i=1}^{n} \left(\frac{\partial F}{\partial x_i} + \sum_{j=1}^{m} \lambda_j \frac{\partial \varphi_j}{\partial x_i} \right) dx_i = \sum_{i=1}^{n} \frac{\partial L}{\partial x_i} dx_i = 0,$$

where $L = F + \sum_{j=1}^{m} \lambda_j \varphi_j$ is the Lagrange function. The functions λ_j should be so that the conditions:

$$\frac{\partial L}{\partial x_i} = 0$$

are fulfilled. Finally, the extremum of the function F is determined by the system of $n + m$ equations given by:

$$\frac{\partial L}{\partial x_i} = 0, \quad \varphi_j(x_i) = 0.$$

Index

Note: Page numbers followed by "f" and "t" indicates figures and tables respectively.

Printed in the United States
By Bookmasters